The Evidence for Evolution

The Evidence for Evolution

Alan R. Rogers

Alan R. Rogers is professor of anthropology and biology at the University of Utah.

The University of Chicago Press, Chicago 60637
The University of Chicago Press, Ltd., London
ⓒ 2011 by The University of Chicago
All rights reserved. Published 2011
Printed in the United States of America

20 19 18 17 16 15 14 13 12 11   2 3 4 5

ISBN-13: 978-0-226-72380-8 (cloth)
ISBN-13: 978-0-226-72382-2 (paper)
ISBN-10: 0-226-72380-1 (cloth)
ISBN-10: 0-226-72382-8 (paper)

Library of Congress Cataloging-in-Publication Data

Rogers, Alan R.
    The evidence for evolution / Alan R. Rogers.
        p. cm.
    Includes bibliographical references and index.
    ISBN-13: 978-0-226-72380-8 (cloth : alk. paper)
    ISBN-10: 0-226-72380-1 (cloth : alk. paper)
    ISBN-13: 978-0-226-72382-2 (pbk. : alk. paper)
    ISBN-10: 0-226-72382-8 (pbk. : alk. paper)
    1. Evolution (Biology)   2. Human evolution.   3. Creationism.   I. Title.
QH361.R64 2011
576.8—dc22

                                                    2010037299

∞ The paper used in this publication meets the minimum requirements of the American National Standard for Information Sciences—Permanence of Paper for Printed Library Materials, ANSI Z39.48–1992.

CONTENTS

1

DARWIN'S
MOCKINGBIRD

The mockingbird hopped out of the bright sunlight and into the shade, then onto the rim of a tortoise-shell cup. It lowered its beak to the water and began very calmly to drink. The cup, as it happened, was resting in the hand of a young naturalist named Charles Darwin, but the bird didn't seem to care. It continued drinking even as Darwin raised the cup for a better look.

Darwin's eyes must have widened with astonishment, but not as much as they might have. These events took place on the Galapagos Islands in 1835. At that time, Darwin was still a creationist and had no way of anticipating the revolution this bird would cause in his own thinking, let alone that of the entire world.

The bird on the cup looked much like the other mockingbirds on the island. Yet these mockingbirds did not look quite like those on a nearby island that Darwin had just visited. And *those* mockingbirds differed from the ones on the next island over. Each island seemed to have its own distinctive mockingbirds. Darwin found this astonishing. The environments offered by these islands were indistinguishable, and the islands were in most cases within sight of each other. Why, Darwin wondered, had the creator made a different mockingbird on each island?

Furthermore, why was the bird on his cup a *mockingbird*? Mockingbirds are found only in the Americas, and Darwin's bird was similar to the ones he had seen in Chile. Yet Darwin was 600 miles from the American mainland. He wondered why the creator had chosen to populate these remote islands with birds that looked so American.

This question was broader than mockingbirds, for the same pattern held for finches and other types of bird. Nor was it just about birds. Each island had its own distinctive tortoises, insects, lizards, and even plants. With only a few exceptions, these were most closely allied to species found in South America.

1

During his stay in the Galapagos, Darwin was able to explain these questions away. He seems to have assumed that the different populations of mockingbird were mere varieties of a single species. This sort of geographic variation is found in many widespread species and would not have challenged Darwin's creationist views. Furthermore, the Galapagos species of mockingbird might have been created in South America and then immigrated to the Galapagos. For all Darwin knew, that species still lived somewhere in South America. Yet within 18 months, this hypothesis came crashing down [104, p. 351]. The difficulties arose after Darwin returned to England, where there were experts on birds, reptiles, insects, plants, and all the other forms of life that Darwin had collected. These experts were eager to examine collections from the far-off Galapagos. In case after case, they assured him that entire species of plant and animal were confined to individual Galapagos islands. Yet the species on different islands were similar to each other and also (to a lesser degree) to South American species. As Darwin put it in 1845,

> one is astonished at the amount of creative force, if such an expression may be used, displayed on these small, barren, and rocky islands; and still more so, at its diverse yet analogous action on points so near each other. I have said that the Galapagos Archipelago might be called a satellite attached to America, but it should rather be called a group of satellites, physically similar, organically distinct, yet intimately related to each other, and all related in a marked, though much lesser degree, to the great American continent. [30, p. 398]

Darwin's solution to this puzzle was subtle. It involved thinking not about the plants and animals that lived on the Galapagos, but about those that did not. There were bats and birds but no native land mammals, reptiles but no amphibians, herbaceous plants but no trees. In each case, the forms that were present were those that seemed best able to survive a long journey across several hundred miles of ocean. Bats and birds can fly, but land mammals cannot. Reptiles and their eggs are resistant to salt water and might have arrived alive on logs after weeks at sea. Amphibians die in salt water and could not survive such a journey. Herbaceous plants have small seeds, which can be carried by wind and in mud on the feet of birds. Trees have larger seeds that cannot travel in this fashion. The Galapagos, it seemed, were populated solely by travelers. This suggested that those plants and animals were not *created* on the Galapagos, but traveled there.

It seemed plausible that these travelers might have come from South America, since that is the closest continent. This accounted nicely for the observation that Galapagos plants and animals were similar to those of South America. But it also raised immediate problems: If they *were* immigrants from South America, why was it impossible to find any Galapagos species in South America? And why were there different species on different islands?

The only explanation, it seemed, was that the immigrants had *changed* after their arrival in the Galapagos. And not only that, the immigrants to each island must have changed once again. This hypothesis would account for all the facts, but it flew in the face of conventional wisdom. For at that time, each species was held to be separately created and unchanging. Darwin's hypothesis was so radical that he did not dare publish it for many years.

During those years, Darwin was hard at work. If you are a good skeptic, you may have noticed some of the same problems that bothered him. Is it really true that the seeds can travel on the feet of birds? How long can the seed of a tree survive in salt water? If Darwin's explanation holds for the Galapagos, then we should find the same pattern in other island chains. Do we? Darwin found ways to answer all these questions and many more. In some cases, his approach was direct and experimental. If we had visited his home during these years, we would have found rows and rows of jars in which seeds soaked in sea water. One wall was hung with ducks' feet, on each of which (if we looked closely) we would have found seeds embedded in dried mud. To answer other questions, he collated information gleaned from the literature and from an extensive correspondence with other scientists. Only after 20 years did he dare to publish. The resulting book—*On the Origin of Species*—is one of the most famous in all of science [27]. In it, Darwin argued not only that evolution happens, but also that the mechanism of evolution is a process he called "natural selection."

Darwin's contemporaries found the first of these arguments more persuasive than the second. During Darwin's lifetime, most working scientists came around to the view that evolution is a fact, but they argued about the importance of natural selection. One hundred and fifty years later, it has turned out that Darwin was essentially right on both counts, but his theory of natural selection left out a lot of details. Those details are still a subject of active research. There is no research, however, about whether evolution happens. That issue was settled over a century ago and is no longer an interesting scientific question.

This has led to a bias in the way we scientists teach courses and write textbooks. We tend to emphasize what we find interesting and to gloss over the rest. For this reason, students learn a lot about the mechanisms of evolution but only a little about the evidence that evolution really happens. (Perhaps this contributes to the fact that most Americans view evolution with skepticism and suspicion.) This book will reverse the traditional emphasis. It will focus on the evidence *that* evolution happens, while saying as little as possible about *how* it happens.

The general structure of the argument is much as it was in 1859. Like Darwin, we must ask: Do species change? Do they split into new species? Does evolution make big changes? Can evolution account for adaptation? These questions form the outline of the book. In every case, however, the answers will involve evidence that Darwin did not have. The case for evolution is stronger today than it has ever been.

2 DO SPECIES CHANGE?

The opinion amongst naturalists that species were independently created, and have not been transmitted one from the other, has been hitherto so general that we might almost call it an axiom.

Thomas Vernon Wollaston, 1860 [56, p. 128]

Breeding experiments on [Drosophila melanogaster] have been going on continuously since 1910; and, as twenty-five successive generations can be bred in the year in the laboratory, some nine hundred successive generations of this fly have been bred. Yet, although many varieties of the fly have been produced, they are all clearly Drosophila melanogaster, *and are all capable of breeding with the parent form—unless so defective as to be incapable of breeding at all.*

Thus, in so far as it is possible to prove a negative, experimental evidence shows that evolutionary theory is not true.

Douglas Dewar, 1949 [36, p. 11]

One hundred and fifty years ago, Thomas Vernon Wollaston found it easy to argue that species did not change (see above). For one thing, no one had ever *seen* one change in the wild. There was plenty of evidence of change in domestic species, but when these escaped into the wild, they seemed to revert to the wild form in a few generations. And judging from the mummified animals found in Egyptian tombs, cats and ibises had remained unchanged for 3,000 years. The fossil record was also disappointing to Darwinists. It did suggest that species occasionally went extinct and were replaced by new species, but it seemed to lack the intermediate forms that (according to Darwin) should link the old to the new. After reviewing all this in 1860, Bishop Samuel Wilberforce concluded that Darwin's new theory was based "on the merest hypothesis, supported by the most unbounded assumptions" [121, p. 248]. The case for fixity of species seemed so strong that Wollaston viewed it as an axiom.

Nearly 90 years later, Douglas Dewar (also quoted above) argued the same position in a published debate with J.B.S. Haldane. In his view, decades of genetic research had produced no evidence that species ever change. But Dewar was on weak ground, for he was ignoring several studies suggesting that species *do* change, as Haldane was quick to point out [36, p. 21].

Today, we know a lot that none of these authors knew. The question of change within species has become especially easy to answer. But this is not the only relevant kind of change. Evolutionists argue that all modern species descend from a single ancestor, and this requires something more: it requires that species sometimes split to form new species. In what follows we take these issues one at a time.

Do species change at all?

It would take a lot of ink to discuss this question in full. We might begin with the fossil record, turn next to variation among domesticated animals, then to results from selection experiments, and finally to direct observation of change within species. But we will discuss only two of these categories of evidence, and those only briefly.

First, there are the fossils. There is evidence of change in any sequence of fossiliferous rocks, but that evidence is often confusing. Most fossiliferous rock accumulates in shallow water near the shore. These deposits are pushed alternately up and down by the same geological forces that cause continental drift (see Chapter 6). As they subside, they accumulate sediment. When they are uplifted, some of that sediment erodes away. The sediment that survives usually represents only a fraction of that originally deposited. The rock record is mostly gaps.

There is one exception to this rule: the sediment of the ocean floor, far from land and far below the surface. Here the sediment accumulates slowly but seldom erodes away. These deposits are studied from core samples, and at first glance they appear devoid of fossils. To find fossils in this rock, the paleontologist works with a microscope.

Core samples from deep-sea sediment often contain the tiny, intricate shells of *radiolarians*. These creatures float near the surface in life, but as they die their shells rain down onto the ocean floor. The shells are pure glass and do not decay. They are common in deep-sea sediment and provide an unusually detailed and continuous fossil record.

6

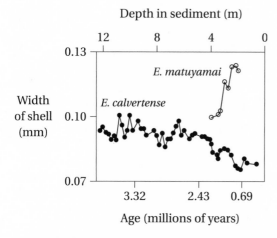

Figure 2.1: Evolutionary change in two radiolarians over about 3 million years [60].

Davida Kellogg and James Hays studied these fossils in the 1970s, and one of their graphs is reproduced in Figure 2.1 [60]. The graph shows change through time in the width of the shell. Early on, the shells of *Eucyrtidium calvertense* (the filled circles) are about a tenth of a millimeter across. They get gradually narrower over the entire three-million-year period. About four meters down, a new species (*E. matuyamai*) suddenly appears. It is possible that we are witnessing here the formation of a new species, as *calvertense* splits in two. This seems plausible because the two species are at first barely distinguishable. On the other hand, the new species may simply have immigrated from another region. However *matuyami* got there, its arrival seems to have provoked rapid evolutionary change in both species. *E. matuyamai* widens rapidly, and *calvertense* narrows even faster than before. As the difference between them grows, the two species are soon easy to tell apart. In cases such as this, there is little room for doubt about the fact of evolutionary change.

Yet some people find such evidence unconvincing. I learned this as a child during a visit to the home of one of my father's brothers in north Texas. That area is full of fossils, and I soon had the driveway covered with them. From a book of Texas fossils I learned that mine all lived in the upper Cretaceous period, between 65 and 100 million years ago. My uncle was skeptical. "Couldn't God have created those rocks all at once," he asked, "with the fossils right in them?" I was at a loss to reply. Years later, I learned that Philip Henry Gosse had made the same argument in the 1850s, and it is easy to find in creationist literature today [47]. For skeptics of this sort, fossils provide no evidence of anything. We'll return to Gosse's argument in Chapter 10. For the moment, let's turn to evidence that's harder to dismiss.

The bacterium *Staphylococcus aureus* usually lives harmlessly in human noses. It can be lethal, however, when it gets under the skin during surgery or as a result of injury. In 1928 Alexander Fleming was searching for a way to fight these infections. His laboratory contained colonies of *S. aureus*, each in a small glass dish. At the end of a vacation, Fleming returned to find his bacterial colonies contaminated with mold. As he surveyed the ruins of his experiments, he noticed something surprising: the mold kept the bacteria from growing. Fleming had discovered penicillin, the first really successful antibiotic.

Penicillin was not used on a large scale until 1943. At that time, resistance was unknown. The first strains of penicillin-resistant *S. aureus* were reported in 1944, and by 1950 the resistant fraction of hospital infections had risen to 40%. This number reached 80% by 1960 and stands at 98% today. As the effectiveness of penicillin declined, doctors turned increasingly to other antibiotics. One of these, methicillin, showed great promise when introduced in 1961. However, the first strain of methicillin-resistant *S. aureus* appeared within a year. This new strain of bacteria, resistant both to penicillin and methicillin, is now common in hospitals. To treat these doubly-resistant bacteria, hospitals now rely on antibiotics such as vancomycin. Resistance to this drug was slow to evolve, presumably because it was less-often used. Recently, however, some strains of *S. aureus* have evolved resistance to it as well [16]. As I write this, the nephew of a friend of mine is struggling with an *S. aureus* lung infection that, according to his doctors, is resistant to all known antibiotics. He is not expected to survive.

There is no question that *S. aureus* has changed, and the genetic basis of these changes is well understood [55]. Furthermore, these changes can be replicated in the laboratory. The typical experiment starts with a colony of *S. aureus* that is not resistant to some drug, say penicillin. This colony is then grown on a medium with a low concentration of penicillin. Because bacterial colonies contain such large numbers of individuals and reproduce so rapidly, it seldom takes long for a variant to arise that confers resistance. Because of its resistance, this variant will reproduce more rapidly. Before long, the entire colony becomes resistant to penicillin. This experiment has been done many times, with various antibiotics [43].

The story of *S. aureus* shows how evolutionary change allows bacteria to adapt to their environments. Similar evidence exists for other organisms. Perhaps the best-documented example in vertebrates is that of the medium ground finch of the Galapagos Islands, which I discuss in Chapter 4. There is little doubt that species change, but this does not prove that they ever change into new species. Indeed, many creationists argue that this latter kind of change never occurs. Let us consider the evidence.

Do species change into new species?

Huxley's challenge

Since ancient times, people have recognized that plants and animals fall into distinct groups, which we call "species." For example, it is easy to tell a lion from a leopard. Not only do they look different, there are (almost) no intermediates. Thus, we group them into separate species.

What keeps species separate? Lions and leopards have been known to mate. If this happened often, and if the resulting hybrid offspring thrived, lions and leopards would eventually blend into a single species. This is clearly not the case, but why?

There is no physical barrier, for lions and leopards live together in the same part of Africa. There are, however, other barriers. First, the two species have never been known to mate in the wild. Even when they do mate, as occasionally happens in captivity, the hybrid offspring have always been sterile. These barriers ensure that genes do not leak from one species into the other: the two species are reproductively isolated and therefore remain distinct.

This idea of reproductive isolation is of central importance. For many biologists, it is a matter of definition: a *species* is a group of organisms reproductively isolated from other groups. But if one defines the word this way, what are we to do about the plant species that hybridize every spring in my wife's garden? What about the fact that 10% of bird species are known to hybridize [49]? For these and other reasons, biologists disagree about how "species" should be defined. No matter what definition we choose, *someone* is sure to disagree. Yet no matter how they define the word, biologists agree on one point: species differences cannot persist between populations that freely exchange genes. There may be some gene exchange—witness the many species that hybridize—but there cannot be too much. If evolution does split species apart, it must do so by curtailing the exchange of genes [68, p. 167]. This suggests that we rephrase our original question. Instead of asking whether species change into new species, let us ask whether evolution can split a population into two parts that are (at least partly) reproductively isolated.

When Darwin published his famous book in 1859, the answer to this question was by no means clear [27]. His friend Thomas Huxley saw this as the weakest point in the argument. He argued that if evolution could *not* produce reproductive isolation, then Darwin's theory would be "utterly shattered" [57, p. 147]. For many years, Huxley's challenge went unanswered, but not for want of trying. Darwin experimented with primroses but was never able to produce a strain reproductively isolated from its parents. In 1922, the geneticist William Bateson [6] was still able to complain that no such evidence had appeared.

But Bateson should have known better, for by then an answer was already at hand. That answer is so strange that no one in Darwin's day could have imagined it. It is the phenomenon of the *polyploid hybrid*.

Polyploid hybrids

In 1899, Frank Garrett was foreman of the Royal Botanic Gardens at Kew, near London. One day he noticed an unusual plant in a tray of primrose seedlings. It was clearly a primrose, but larger and with unusual leaves. When it flowered, the new primrose was magnificent. It won a First Class Certificate at the meeting of the Royal Horticultural Society in 1900. Garrett naturally wanted more of these lovely primroses, but there was a problem: the plant was sterile, producing neither pollen nor seeds. Garrett had a hunch that the new plant was a hybrid, so he tried crossing the most likely parental species. Sure enough, the hybrid offspring looked like the prize-winning plant. Garrett called his new hybrid the Kew primrose (*Primula kewensis*). The new plants were as beautiful as the original but just as sterile.

Things rested there until 1905, when another gardener with sharp eyes made another discovery. One branch of one plant had begun making fertile flowers. These yielded seeds, which grew into Kew primroses. But these primroses were fertile, producing pollen and seeds like any normal plant. The new primroses could reproduce with each other, but not with either of the two parental species. A completely new species of primrose had appeared under the eyes of these watchful gardeners.

No one knew how this had happened. Why was the original Kew primrose sterile? Why were the new ones fertile? These questions attracted the interest of a young woman named Lettice Digby, a botanist at the Royal College of Science. Peering into a microscope, she counted 18 strands of DNA—18 *chromosomes*—in each cell of the original sterile Kew primrose. This was no surprise, for the two parental species also carried 18 chromosomes. The surprise came when she examined cells of the fertile hybrid. Instead of 18 chromosomes, these cells carried 36. The Kew primrose had become fertile by doubling the number of its chromosomes. Digby had documented the first example of what we now call a *polyploid hybrid* [38].

Within a few years, it was clear how they worked. Like any other product of sexual reproduction, the initial sterile primrose got one copy of each chromosome from its mother and one from its father. During sexual reproduction, the chromosomes must line up in pairs, and this was where the problem arose. The chromosomes in the sterile primrose came from different species and were too different to pair properly. There was no such problem in plants with doubled chromosomes—these had two copies of each chromosome, which could pair without difficulty.

Botanists soon developed tricks to encourage a plant to double its chromosomes, and these have proved useful in agriculture. Many of the vegetables and flowers in our supermarkets and florist shops are new species created by polyploid hybridization.

In view of all these new species, it is no longer possible to argue (as people did in Darwin's day) that the number of species was fixed at the moment of creation and has not changed since. This had become clear by the mid-1920s, but it was not enough to satisfy Arne Müntzing, who studied plant genetics in Sweden. The problem, as Müntzing saw it, was that the new hybrids had all been developed by people. It was tempting to *assume* that the same thing happened in nature, but how could one really know?

Müntzing had been working with several related plants called hempnettles. Two of these (*Galeopsis speciosa* and *G. pubescens*) carry 8 chromosomes in each pollen cell, whereas a third (*Galeopsis tetrahit*) carries 16. Müntzing suspected that *tetrahit* was a natural polyploid species derived from an ancient hybridization of *pubescens* and *speciosa*. He set out to test this hypothesis.

In the late 1920s, he began hybridizing the two supposedly ancestral species. The initial hybrids were nearly sterile, but Müntzing was patient. He eventually got one of these plants to produce a single seed in which the chromosome count had doubled. As he had predicted, this seed's progeny looked just like wild *Galeopsis tetrahit*. Furthermore, they could *breed* with wild *tetrahit*. They could not, however, breed with either parental species. Müntzing had reconstituted a wild species. His experiment proved that new species form by hybridization not only in the laboratory, but also in nature [73].

Since Müntzing's day, research has shown that many plant species arise as polyploid hybrids [23]. Yet Huxley would probably not be entirely satisfied. Many species of plants and nearly all species of animals have arisen *without* doubling their chromosomes. For this reason, Huxley would point out that polyploid hybridization is not enough. Darwin's theory fails unless new species can also arise in some other way. Where, he would ask, is the evidence?

Artificial selection experiments

In the 1940s when Karl Koopman started graduate school at Columbia University, Darwin's famous argument lay nearly 90 years in the past. It was clear that some species had arisen as polyploid hybrids, but it was also clear that most had not. Biologists were convinced that selection could split a species in two, but the evidence was indirect. No one had ever seen it happen in an experiment. Koopman set out to change this.

His real passion was for mammals, and he would go on to a long career studying bats. Yet these were hardly ideal for the experiment that he had in mind. His experiment might require many generations. With mammals, that could take years, and Koopman might have grown old and gray in graduate school. He decided to work instead with fruit flies, the tiny insects that you may have seen hovering around overripe bananas. Fruit flies are easy to raise (all you need are bananas), and their generations take only a matter of weeks.

Koopman's experiment involved a method called *artificial selection,* which has been used for centuries to improve domestic plants and animals. Each generation, the breeder selects individuals that excel in some way—the cows that yield the most milk or the tomato plants with the largest fruit—as parents of the next generation. After generations of selection, the breeder gets an improved variety of milk cows (or tomatoes). Koopman used the same method for a different purpose.

He was trying to show that selection could reduce the rate of interbreeding. Thus, he started with two populations of flies. Both types of fly preferred to mate with others of their own type, but some were less choosy than others. In any mixed collection of flies, there would be some interbreeding. Koopman simply discarded the resulting hybrid offspring. The only flies that reproduced successfully were those that avoided interbreeding.

As the generations passed, Koopman's flies became more choosy about mates, and the rate of interbreeding declined. This experiment proved that selection *can* reduce the rate at which populations exchange genes. And as we discussed above, this is all one needs to do to produce a new species. Yet Koopman's experiment was not entirely convincing, because his two initial populations came not from the same species but from two closely related species. He had not split a species in two; he had merely reduced the rate of hybridization between two existing species.

Since that time, biologists have performed many similar experiments. Most have involved two populations from the same species and thus avoided the weakness in Koopman's experiment. Yet they generally obtain similar results: when one selects against interbreeding, the rate of interbreeding usually declines [93].

These experiments demonstrate that selection can reduce the rate at which populations exchange genes. Yet it is hard to be sure they have really produced separate species. The trouble is that these experiments usually do *not* reduce the rate of gene exchange all the way to zero. There is usually some interbreeding even at the end. Presumably, this reflects the fact that isolating mechanisms arise only gradually. In nature, we find some isolation between populations within species, and some new species—only a few million years old—still hybridize occasionally [49, 52, Chapter 16, 110]. Biologists are never able to run experiments for thousands of generations, so it is hardly surprising that few experiments reduce hybridization all the way to zero.

This is no fatal flaw, for many natural species hybridize too. As discussed above, species remain distinct unless they hybridize *too much*. How much is too much? To answer this question, we would need to adopt a particular definition of "species" and then dive into mathematical theory. These are both issues I would like to avoid, so let us take the easy way out. Let us ask whether selection can reduce the rate of hybridization *all the way to zero*. If it can do this, it can clearly split a species in two.

To find such an experiment, we turn to the work of Bill Rice and George Salt [94]. They pointed out that in nature, fruit flies search in a complex environment for places to mate and reproduce. Those that converge on the overripe banana on your kitchen counter have made a series of similar decisions: Search high or search low? In light or in shade? Prefer the smell of fresh fruit or of rotting fruit? And so on. Flies that made different decisions ended up in different places and did not mate with the ones on your counter. This suggests that selection can produce reproductive isolation simply by tweaking the way fruit flies make decisions. Rice and Salt set out to test this idea in an experiment.

Their experimental apparatus was a twisted three-dimensional maze of tubing. In each generation, all flies emerged as adults in the middle of the maze and then made their way through the twisted tubing, making a series of decisions (up/down, light/dark, etc.) along the way. Depending on their decisions, they ended up in one of eight habitats. Six of the eight habitats were dead ends—flies that went there were discarded. All the action was in the remaining two habitats, and life was dangerous even there, at least for female flies. Those who chose a different habitat than their parents faced oblivion. The only female flies that got to breed were those that chose the same habitat as their parents. The results are shown in Figure 2.2. In the early generations, flies seem to have chosen habitats at random. By the end of the experiment, the population had split completely in two. Every single fly chose the same habitat as its parents.

Because of their consistent preference for habitat, the flies in the two habitats had become completely isolated. There was no longer any need for Rice and Salt to cull female flies who switched habitats, because there were no such flies. None of the male flies switched either. In the absence of movement from one habitat to the other, the two populations could not exchange genes. They

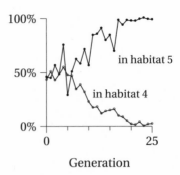

Figure 2.2: Percentage of flies breeding in habitats 4 (∘) and 5 (•) that were hatched in habitat 5, in experiment of Rice and Salt [94]. Isolation was eventually complete because all flies chose the habitat in which they hatched.

were in effect separate species. Rice and Salt had proved that selection can split a species in two.

These experiments provide the experimental evidence that Huxley had demanded. They show that selection can split a species into completely isolated parts. Perhaps the split between lions and leopards evolved in a similar way—the result of a subtle shift in habitat preference between two ancient populations of cats. Müntzing, however, would probably remain unconvinced. He would object—as he did about the early experiments on polyploid hybrids—that experiments only tell us what happens in the laboratory. They do not prove that selection ever erects barriers between species in nature.

To meet this objection, we need to study millions of individuals over millions of years. This is beyond the scope of any laboratory experiment, so we must turn to natural experiments, such as *ring species*.

Ring species

The tiny kingdom of Bhutan clings to the Himalayan slopes that form the southern edge of the Tibetan Plateau. If you were to visit Bhutan during the warmer months and hike up to where the trees begin to thin, you might find the small bird that English speakers call the greenish warbler (*Phylloscopus trochiloides*, Figure 2.3). With luck, you might hear one sing. Although the song is pretty, it is very simple, sounding something like "sweety sweety sweety sweety."

Should you take the impossibly long hike west and north along the edge of the Tibetan Plateau, you would find greenish warblers all along the way. You would find them in Nepal, in Kashmir, and right across the mountains into Tajikistan, Kyrgyzstan, and Siberia. All along this route, the birds change by imperceptible degrees. You might not notice, for each day they would look and

Figure 2.3: The greenish warbler, *Phylloscopus trochiloides.* Artwork copyright © 2011 Jude Higgins.

sound like those the day before. But by the time you reached Siberia, the birds would be 10% smaller, and their songs would not sound like those in Bhutan. The greenish warblers of Siberia sing complex songs that I will not even try to represent with words.

You could also reach Siberia by walking the opposite way around the Tibetan Plateau: east from Bhutan, then north through China and Mongolia. The result would be much the same: the birds would get gradually smaller and their songs more complex. The surprise emerges in Siberia, where the two routes converge. The eastern and western populations of greenish warbler overlap there, and both can be heard singing in the same areas. Both groups sing complex songs, but these songs are different. Furthermore, the two groups do not interbreed—they behave as separate species. Yet they do interbreed all along both routes north from Bhutan. This peculiar pattern was first noticed in the 1930s and has recently been confirmed with genetic data [58].

Because of their circular pattern of variation, such species are called *ring species*. At present, all greenish warblers are connected by gene flow, so there is only one species. Differences among greenish warblers are like those that evolve in any other widespread species. Yet if the southern populations went extinct, the eastern and western populations would then be separate species.

Many creationists accept the reality of evolutionary change within species but deny that evolution can split a species in two. Ring species make this view untenable. They prove that the evolutionary processes that generate variation within species can also separate one species from another. New species evolve not only in the laboratory, but also in nature.

Conclusion

This chapter has shown that evolution not only produces changes within species but also creates new ones. Both conclusions rest on direct evidence: We can *see* species changing. We can also see new species forming by polyploid hybridization and by artificial selection. There is excellent evidence that this happens in nature as well as in the laboratory.

Nonetheless, there are still skeptics. For example, some authors accept that evolution produces new species but argue that it does not produce major taxa such as amphibians, reptiles, mammals, and birds. According to this argument, evolution makes minor adjustments but not major changes. To address this argument, we will need different evidence.

3   Does evolution make big changes?

We have established that evolution makes small changes, but does it also make large ones? Did it make amphibians from fish, and mammals from reptiles? Evolutionists think so, but consider another possibility. Perhaps God created many kinds of plants and animals—all the major taxa—and then let them evolve. Perhaps evolution makes only small changes, and these never take a species far from its original form. If so, then leopards, lions, and domestic cats may all have evolved from some ancestral cat, but no mammal would share a common ancestor with any reptile. Life would have not one origin but many.

This is an old idea now called *progressive creationism* [40, pp. 383–384]. Although it was originally promoted as a way to reconcile science and scripture [75], modern creationists often give it a secular rationale. For example, Michael Denton [34] argues that each major taxon is separated from all others by maladaptive intermediates. For example, a species intermediate between fish and amphibian would fare poorly and tend to die out. Consequently, the major taxa cannot have evolved from a common ancestor.

One of Denton's examples involved whales and porpoises. These creatures are entirely aquatic, with no visible hind limbs. Yet if all mammals had a single origin, then some ancestral whale must have walked on land. But how could land mammals evolve into whales? What sequence of small adaptive changes could possibly lead from the one to the other? Denton argued that whales cannot have evolved, because no such sequence is possible. This idea has a long history. Here is how Douglas Dewar put it 50 years earlier:

> The gradual transformation of a land animal into a whale or a sea-cow appears to be physically impossible, because the tail could not act as a propeller by vertical motion until the pelvis had been so reduced in size as to render locomotion on land impossible. If such

17

> transformation occurred gradually there must have been a long
> period when the ancestors of these aquatic forms, while yet poor
> swimmers, were unable to use the hind limbs for locomotion. How
> could such creatures hold their own in the struggle for existence?
> [35, p. 61]

Dewar and Denton were unable to fathom how whales could have evolved from
land mammals. From this, they concluded that it didn't happen. This is what
Richard Dawkins calls the "argument from personal incredulity" [31, p. 38].

But as you will soon see, incredulity is an unreliable guide.

Fossils of intermediate forms

If whales evolved from land mammals, there must have been intermediate
animals—neither fully terrestrial nor fully aquatic. With luck, we might find
their fossilized remains. By the same argument, there must also have lived
many other intermediate forms: animals that combined the characteristics of
fish and amphibians, of reptiles and birds, and so on. Any of these might show
up in the fossil record. Yet in Darwin's day, none had. You can still sense the
frustration in his remark:

> Why then is not every geological formation and every stratum full
> of such intermediate links? Geology assuredly does not reveal any
> such finely graduated organic chain; and this, perhaps, is the most
> obvious and gravest objection which can be urged against my the-
> ory. [27, p. 280]

In 1859, Darwin could only suggest that the fossil record was incomplete. He
predicted that the missing intermediates would eventually show up, and this is
exactly what has happened. The record is especially good for whales.

If you could travel back in time 50 million years, you would find India separated from Eurasia by a shallow sea. Along the shore, in what is now Pakistan,
you might encounter the animal that modern paleontologists have named
Pakicetus (Figure 3.1). It had the teeth of a carnivore and walked on all fours
like—say—a medium-sized dog. From a distance you might even have mistaken it for one. Up close, however, you would notice something strange about
the toes: they ended not in claws, but in small hooves. In fact, the bones of
its ankle identify this animal as an artiodactyl—a member of the order that
includes pigs, sheep, deer, camels, and hippopotamuses. In a host of other
anatomical details, however, it was unlike anything that now walks on land.
Pakicetus, you see, was a whale.

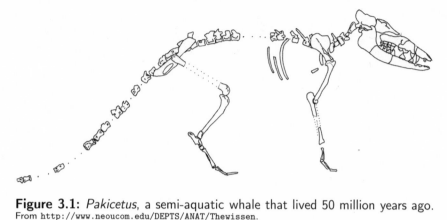

Figure 3.1: *Pakicetus*, a semi-aquatic whale that lived 50 million years ago.
From http://www.neoucom.edu/DEPTS/ANAT/Thewissen.

To understand this remarkable claim, we must first ask what *Pakicetus* did for a living. Although it looked more or less like a terrestrial carnivore, it probably did not outrun its prey. The bones of its legs were so dense and heavy that it would have been a slow runner. These dense legs might however have been useful as ballast, making it easy to lurk just below the water's surface. It also had long toes, which would have helped it swim. These anatomical details, along with others that we will get to in a moment, suggest that *Pakicetus* spent a lot of time in the water. But on the other hand, this animal *looked* like a terrestrial carnivore. Until very recently, paleontologists could not decide how *Pakicetus* made a living.

The decisive evidence came from the atoms of which *Pakicetus*'s skeleton was made. The atoms of a given element are not identical. They vary slightly in weight, or mass, and these variant forms are called "*isotopes*." The bones of aquatic and terrestrial animals differ in their isotopes, and these differences persist even in fossils. In *Pakicetus*, the isotopes of oxygen and carbon display the characteristic signature of an animal that spent most of its time submerged in fresh water [20].

Living as it did on the interface between land and water, *Pakicetus* had to cope with an unfortunate fact: ears that hear well in air tend to hear poorly under water. If you have ever been swimming, you will know about this. For humans, the underwater world is a world of silence, even though sound travels better through water than air. Like us, *Pakicetus* could hear well in air. But under water its world probably did not fall quite so silent. Its underwater hearing benefited from a series of modifications to the anatomy of its ear [81]. It is these adaptations, more than any others, that identify *Pakicetus* as a whale.

19

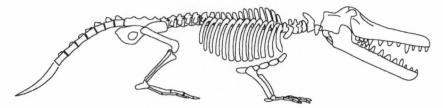

Figure 3.2: *Ambulocetus*, an amphibious whale that lived 49 million years ago. Please refer to the original publication: J.G.M. Thewissen, et al. Ambulocetus natans, *an Eocene cetacean* (Mammalia) *from Pakistan*, E Courier Forschungs-Institut Senckenberg Stuttgart, 1996, by permission of Senckenberg forschungsinstitut und naturmuseum.

Not that this ear was closely similar to that of any modern whale. It exhibits only the first few evolutionary steps in that direction. But the fossil record of whales is now so good that we can trace the progress of that ear, step by step, from the primitive condition found in *Pakicetus* to the sophisticated ear of a modern whale [81].

A million years later, the swamps and rivers of Pakistan were home to a much larger animal, *Ambulocetus* [107] (Figure 3.2). You would not mistake this animal for a dog, but you might mistake it for a crocodile, as it floated nearly submerged. If you got too close, the result would have been much the same, for *Ambulocetus* was 12 feet long and would have regarded you as prey. But this animal was no crocodile. Like *Pakicetus*, it had the hoofed toes of an artiodactyl and the ear bones of a whale. It could not however run as well as *Pakicetus*: its legs were short, and its feet—especially the hind feet—were very large. These provided most of the propulsive force when it swam and also when it lunged forward after prey. *Ambulocetus* was an amphibious whale.

A few million years later, Pakistan was home to an even more aquatic whale, *Rodhocetus* [44] (Figure 3.3). Its legs were shorter still, and it was awkward on land. It still paddled with its hind feet, and the muscle attachments on the sides of its toes indicate webbed feet. But these were probably of secondary importance, for the real propulsive force came from the tail. *Rodhocetus* had a flexible spine with muscle attachments for a powerful tail, which provided most of the force that pushed *Rodhocetus* through the water. *Rodhocetus* was more aquatic than any previous whale, but it still had the hooves and ankles of an artiodactyl along with the skull, teeth, and ear bones of a whale [45].

Over the next few million years, whales relied less and less on their legs. By 40 million years ago, we find creatures like *Dorudon* (Figure 3.4), which resemble large modern porpoises. They were about 15 feet long and torpedo-shaped, with flippers for forelimbs and powerful tails for propulsion. However, these

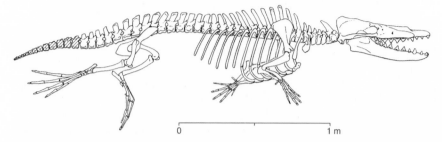

Figure 3.3: *Rodhocetus*, an amphibious whale that lived 46–47 million years ago. The tail was probably much longer than shown here. From P.D. Gingerich, et al. Origin of whales from early artiodactyls: Hands and feet of eocene protocetidae from Pakistan. *Science*, 293:2239–2242, 2001. Reprinted with permission from AAAS.

Figure 3.4: *Dorudon*, an aquatic whale that lived 36–40 million years ago. From P.D. Gingerich, et al. New protocetid whale from the middle Eocene of Pakistan: Birth on land, precocial development, and sexual dimorphism. *PLoS ONE*, 4(2):e4366, 2009.

porpoises had functional hind legs, complete with feet and toes. Very small legs, it is true—they could not have supported *Dorudon*'s weight. They do, however, show clearly that *Dorudon* had terrestrial ancestors. These animals were fully aquatic and no longer restricted to the coasts of Pakistan. They have been found both in Egypt and South Carolina; they probably lived throughout the tropical oceans.

From *Dorudon*, the evolutionary path to modern whales is straightforward. Hind limbs dwindled. Some modern whales have lost them altogether, whereas others retain a tiny vestigial pelvis and femur deep inside. Brains got larger, and whales evolved an elaborate mechanism for "seeing" underwater with ultrasound. But the fundamental adaptations that make them aquatic were already in place by the time of *Dorudon*.

Each of these fossil species has more aquatic adaptations than the last. We see this, for example, in the joints of the elbow, wrist, and fingers. These were still mobile in *Ambulocetus*, like those of a land mammal. In *Dorudon*, the fingers were still mobile, but the wrist and elbow had become stiff, like those of modern whales. The nasal opening moved gradually from a position near the tip of the snout in *Pakicetus* to one closer to the eyes in *Dorudon*. In modern whales, it is even farther back [107].

It is hard to imagine a better sequence of intermediate forms, but there are others just as good. Figure 3.5 shows several fossil species that bridge the gap between fish and amphibians. The bottom one, *Tiktaalik* had fins that ended in rays like any normal fish. But behind those rays, there were a wrist, an elbow, and a shoulder. This fish could bend its fin at the wrist and walk in shallow water. Ten million years later, *Acanthostega* had fingers (eight of them!) instead of fin rays, but still had the tail of a fish [3, 126]. *Ichthyostega* looks more like an amphibian still but still has fin rays in its tail.

Figure 3.6 shows a drawing of another transitional fossil, which has recently settled a dispute that goes back to the 1860s. The dispute involves the evolution of "flatfish," such as flounder, plaice, sole, and halibut. These fish spend their days lying on their sides on the sandy bottom. If their eyes were arranged like those of normal fish, one eye would be pressed uselessly against the sand. But that is not the case, for in these fish both eyes are on the same side. The question is, how did this condition arise? St. George Mivart was the first to pose the question. He observed in 1869 that it was far from clear "how the transit of one eye a minute fraction of the journey towards the other side of the head could benefit the individual" [72]. A small change in the position of the eye would still leave that eye pressed uselessly against the sand, so how could such changes be favored by selection? And if they could not, then how could the asymmetrical

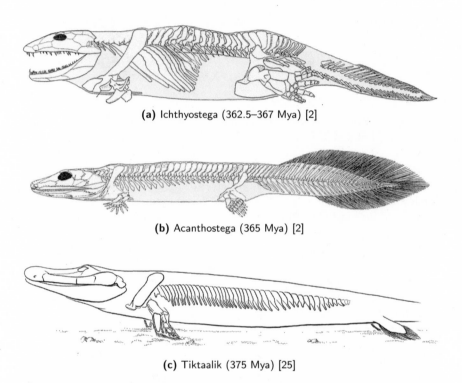

(a) Ichthyostega (362.5–367 Mya) [2]

(b) Acanthostega (365 Mya) [2]

(c) Tiktaalik (375 Mya) [25]

Figure 3.5: Intermediates between fish and amphibians.

[a] Reprinted by permission from Macmillan Publishers Ltd: *Nature* (P.E. Ahlberg, J.A. Clack, and H. Blom. The axial skeleton of the Devonian tetrapod *Ichthyostega*. *Nature*, 437(7055):137–140, 2005), ©2005. [b] Reprinted by permission of P.E. Ahlberg and J.A. Clack. [c] Reprinted by permission from Macmillan Publishers Ltd: *Nature* (E.B. Daeschler, et al. A Devonian tetrapod-like fish and the evolution of the tetrapod body plan. *Nature*, 440:757–763, 2006), ©(2006).

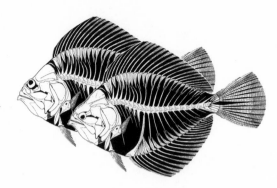

Figure 3.6: Eocene flat-fish *Amphistium*. We see the upper surface of the individual in front and the lower surface of the one behind. Reprinted by permission from Macmillan Publishers Ltd: *Nature* (Matt Friedman. The evolutionary origin of flatfish asymmetry. *Nature*, 454:209–212, 2008), © (2008).

eyes of flatfish have evolved by natural selection? As usual, Darwin had an answer. He pointed out that young flatfish are like normal fish, with eyes on either side. Even in this symmetrical condition, the young fish spends some time lying on its side. Its lower eye is not however totally blind, because of the fish's flexible skeleton. This flexibility enables the fish to bend its head and raise the lower eye off of the sand [29, pp. 186–188]. It is easy to imagine that small changes in the position of the lower eye might have been beneficial because these would have allowed the young flatfish to get this eye farther from the sand. Nonetheless, Mivart's argument has been echoed by creationists of every generation since.

In July 2008, paleontologist Matt Friedman laid this question to rest. He studied fossils of two flatfish genera that lived 45 million years ago and showed that although their eyes are asymmetrical, they are still on opposite sides of the head. Figure 3.6 illustrates one of these genera, *Amphistium*. Note that the eyes of these fish are in different positions. That of the rear fish is higher on its head. This is because the rear fish lay left-side down, and the front fish right-side down. When these fish lay flat on the sand, their upper eyes were in a normal position, but their lower eyes had migrated part (but only part) of the way to the upper side. This intermediate fossil proves that the peculiar asymmetry of flatfish eyes evolved gradually, as Darwin proposed. It did not arise suddenly, as Mivart claimed it must have.

The first really convincing transitional fossil was found in 1861, soon after the publication of *Origin of Species*. *Archaeopteryx* was an animal that lived among dinosaurs and looked a bit like a long-legged crow. It had wings with feathers and could almost certainly fly. Under the skin, it had the wish bone of a modern bird. But instead of a beak, it had jaws with teeth, and it carried a long reptilian tail. Its foot was also like that of a dinosaur rather than that of a bird.

It is hard to decide whether to group *Archaeopteryx* with birds or dinosaurs. Many dinosaurs had feathers, and some had wish bones. All in all, *Archaeopteryx* was a confusing mixture of bird and dinosaur [120].

Where does this leave progressive creationism? If all major taxa had been separately created, then there should be *no* intermediate fossils. This is clearly not the case. On the other hand, no one would argue that intermediate fossils are common. Why should they be so rare?

The history of whale paleontology offers a clue. Paleontologists have known about fossil whales for a century and a half, but they only recently discovered the terrestrial and amphibious whales discussed here. It took a long time to find them because they lived only in the vicinity of Pakistan, and the whole transition from land to sea was over within 10 million years. To find transitional whales, you had to look in Pakistan and in rock of just the right age. If that rock had lain deeply buried or had been lost to erosion, we would never have known about these whales. It is no wonder that intermediate fossils are rare.

Even if the rarity of intermediate fossils is unsurprising, it is still unfortunate. If they were more common, the fossil record might document all the major transitions of evolutionary history rather than just a few. What of the undocumented transitions? How can we be sure they really happened? The answer, as we shall see, lies not in fossils but in things that are still alive today.

Traces of common descent in living organisms

Evolutionists and progressive creationists agree about a lot. They agree, for example, that any group of closely related species shares a long evolutionary history—a long history of changing and splitting. Furthermore, they agree that within such a group all species descend from a single ancestor. The relationships among such species must therefore have the form of a tree.

A tree of relationships among species is called a phylogeny. Figure 3.7 shows the phylogeny of several hypothetical species of butterfly. The white ancestral species splits into two daughter species, one of which then evolves gray coloration. This gray species then splits to form two daughter species of its own. Both daughters share the gray coloration of their parent, a reflection of shared ancestry.

This sort of tree is consistent not only with the views of evolutionists, but also with those of progressive creationists. The dispute between them is not about the reality of phylogenetic trees; it is about the number of ancestors. For a progressive creationist, there are lots of independently created ancestors and

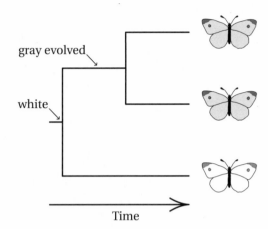

Figure 3.7: Hypothetical phylogenetic tree.

thus many independent trees. For an evolutionist, these trees are not independent; they are branches in a larger tree that unites all living things. Either way, closely related species share ancestors.

And because they share ancestors, they should also share similar values of many characters. We should not, however, expect the same pattern in all characters, for it can be spoiled in various ways. For example, one of the gray butterflies might revert to the ancestral color, white. The two butterflies most similar in color would then be the *least* closely related. Such cases are rare because they require multiple evolutionary changes. They do however exist. The trick with phylogenetic trees is to avoid being fooled by these misleading cases.

There is one type of character that avoids these difficulties. One so easy to use that any school child can infer a phylogenetic tree, simply by inspecting the data. Because few non-geneticists have ever heard of it, a little background on DNA is required.

We often think of DNA as the collection of blueprints from which our cells make proteins. These blueprints, called "*genes,*" are passed from parent to child and account for the resemblance between parent and offspring. Yet they constitute only a small fraction—a few percent—of our DNA. Some of the rest is used in other ways, but most has no known function. To the best of our knowledge, it is nothing more than molecular junk.

Some of this junk DNA is worse than useless: it consists of molecular parasites. These parasites are shortish stretches of DNA that do nothing useful for you and are good at just one thing. They have evolved the ability to hijack the machinery of your cell, tricking it into copying the parasite and inserting the copy into some random spot in your DNA. Your DNA is riddled with these

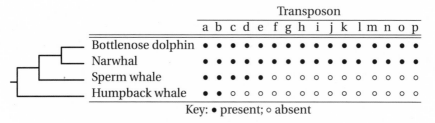

		Transposon														
	a	b	c	d	e	f	g	h	i	j	k	l	m	n	o	p
Bottlenose dolphin	●	●	●	●	●	●	●	●	●	●	●	●	●	●	●	●
Narwhal			●	●	●	●	●	●	●	●	●	●	●	●	●	●
Sperm whale	●	●	●	●	●	○	○	○	○	○	○	○	○	○	○	○
Humpback whale	●	●	○	○	○	○	○	○	○	○	○	○	○	○	○	○

Key: ● present; ○ absent

Figure 3.8: Transposon data and corresponding tree for whales [77].

parasites and copies of parasites. Each one is called a *transposon*. In popular literature, transposons are known as "jumping genes."

Although they do you no good, most transposons are not really harmful either. And they are profoundly useful to those of us who study phylogenetic trees. What makes them so useful? In the first place, it is extremely unlikely that two transposons will ever insert at the same spot. This means that two individuals who share the same transposon at the same spot in their DNA must also share an ancestor. It is also likely that *all* the descendants of this ancestor will carry the transposon, because transposons are seldom lost. They do eventually degrade and become impossible to identify. But they do not lie even then; they are simply uninformative. These uninformative cases will appear as question marks in our data.

Because of these advantages, Norihiro Okada used transposons to study the phylogeny of whales and porpoises. He and his colleagues found 25 transposons in the DNA of these animals and used them to look at 14 species. A subset of their data is shown in Figure 3.8. Each column there represents a different transposon, and the symbols indicate whether that transposon is present (●) or absent (○) in each species. The table shows that the first two transposons are present in all four species. To explain this fact, a creationist explanation would require that a transposon insert in exactly the same spot in the DNA of all four species. It is exceedingly unlikely that this could happen by chance in even two species. Yet the creationist must assume that it happens in all four, an improbability of breathtaking proportion. And since there are two such transposons, the creationist needs two breathtaking improbabilities, just for the first two transposons. The obvious alternative is that these transposons inserted in the DNA of the common ancestor of all four whales.

Notice that the ● and ○ symbols are not scattered randomly but occur in blocks. We have already discussed the first block on the left (columns a–b). The transposons of the next block (columns c–e) occur in all species except

							Transposon									
	a	b	c	d	e	f	g	h	i	j	k	l	m	n	o	p
Bottlenose dolphin	•	•	•	•	o	•	o	•	o	•	•	o	o	•	•	•
Narwhal		•	•	•	•	•	•	•	o	o	o	o	o	o	o	•
Sperm whale	•	•	•	o	•	o	•	o	•	•	•	•	•	o	o	
Humpback whale	•	•	o	•	•	o	o	o	•	o	o	•	•	o	•	o

Key: • present; o absent

Figure 3.9: Whale transposon data after shuffling the symbols within each column. These data are not consistent with any tree.

the humpback. These species must share a common ancestor not shared by the humpback. In other words, they form a separate branch of the tree. The remaining transposons occur only in the bottlenose and the narwhal, which must therefore form their own sub-branch. The tree implied by these data is also shown in the figure. The different transposons tell a consistent story and indicate that whales share a long history of changing and splitting that began with a single ancestor.

We can be quite sure that the pattern in these data did not arise by chance. There are two reasons for this, the first of which was discussed above. It is exceedingly unlikely that a transposon would occur by chance at the same spot in the DNA of several species. Yet this happened with every one of the transposons in Figure 3.8. And even if we accept all these multiple insertions, there is yet another source of improbability.

If the species had been independently created, there would be no reason to expect the • and o symbols to be arranged in a pattern that implies a phylogenetic tree. This might happen by chance, but how likely is that? To find out, let us temporarily accept that transposons a–b are present in all four species, that c–e were present in three of the four, and so on, just as in Figure 3.8. We can generate "shuffled" data sets by randomly re-ordering the • and o symbols in each column of Figure 3.8. One such data set is shown in Figure 3.9. Take a moment to examine it, and ask yourself what tree it implies. You will soon discover that it isn't consistent with *any* tree. For example, transposon c implies that the humpback split off first from the other species, yet transposon d implies that the sperm whale split off first. These cannot both be true. I programmed a computer to generate shuffled data sets. In 100,000 tries, it didn't find a single shuffled data set that would fit onto a tree—*any* tree. Thus, the perfect phylogenetic pattern in Figure 3.8 is extremely unlikely to arise by chance.

So far, the results are consistent not only with evolution but also with progressive creationism. Whales are a group of closely related species. Within such

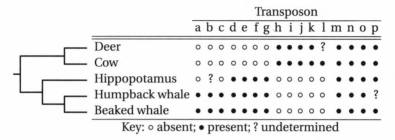

	Transposon															
	a	b	c	d	e	f	g	h	i	j	k	l	m	n	o	p
Deer	○	○	○	○	○	○	○	●	●	●	?	●	●	●	●	●
Cow	○	○	○	○	○	○	○	●	●	●	●	●	●	●	●	●
Hippopotamus	○	?	○	●	●	●	●	○	○	○	○	○	●	●	●	●
Humpback whale	●	●	●	●	●	●	●	○	○	○	○	○	●	●	●	?
Beaked whale	●	●	●	●	●	●	●	○	○	○	○	○	●	●	●	●

Key: ○ absent; ● present; ? undetermined

Figure 3.10: Transposon data and corresponding tree for whales and relatives [78].

groups, both theories predict a phylogenetic pattern of relationships. Let us now consider a more controversial problem: the relationship of whales to other mammals. If whales were created separately, there is no reason to expect phylogenetic pattern in this broader comparison.

Okada studied this problem too, and some of his data are shown in Figure 3.10. Now we have very different kinds of mammal: whale, hippopotamus, cow, and deer. This new table is also organized in blocks, which hint at an underlying phylogenetic pattern. To understand this pattern, take the blocks one at a time just as we did above. You should end up with the tree shown in the figure. As before, the phylogenetic pattern in these data is unlikely to have arisen by chance. (I examined 100,000 shuffled data sets, not one of which would fit onto any tree.) These data make sense if whales, hippos, cows, and deer all evolved from a single ancestor. They are hard to explain any other way.

The molecular data also agree with the fossils. The molecular tree (Figure 3.10) shows that whales are closely related to the hippopotamus, an artiodactyl. Indeed, the hippo is more closely related to whales than to other artiodactyls (deer and cow). This shows that the ancestral whale was indeed an artiodactyl, just as we concluded earlier from fossils.

Agreement between fossil and molecular data exists all the way up the vertebrate family tree. Figure 3.11 shows two phylogenetic trees for the major vertebrate taxa. The one on the left was inferred from morphology, the one on the right from 118 gene families. The two trees are nearly identical, just as predicted by common descent.

A creationist might argue that there is no mystery about the agreement between the morphological and genetic patterns. Morphological characters are after all *shaped* by the genes, so morphological similarities ought to reflect genetic similarities. This might account for the similarity of these patterns, but

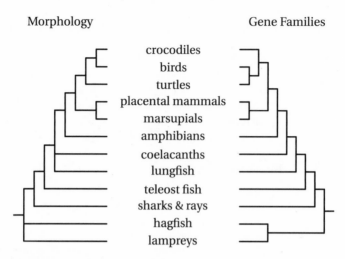

Morphology   Gene Families

crocodiles
birds
turtles
placental mammals
marsupials
amphibians
coelacanths
lungfish
teleost fish
sharks & rays
hagfish
lampreys

Figure 3.11: Vertebrate phylogenetic trees from morphology and 118 families of protein-coding genes [21].

it does not explain why both have the form of a tree. Common ancestry, on the other hand, explains it all.

Traces of common ancestry are especially striking in developmental genes, many of which play the same role in distantly related species. For example, the gene *pax6* induces an embryo to develop eyes, and it does so in an amazing variety of animals, including vertebrates, insects, segmented worms, mollusks, and flatworms [14, p. 122]. It must have acquired this role in the common ancestor of all these taxa.

One cannot trace the phylogenetic tree all the way back to the origin of life. The farther back we go in time, the more uncertain our estimates become. It may be that trees are the wrong way to think about this earliest history. In life's earliest stages, the boundaries between species may have been blurry; early species may have swapped genes more promiscuously than modern ones do.

In spite of this uncertainty, there are several good reasons to think that life had a common origin. Central among these is DNA itself. A DNA sequence looks like gibberish, but any of your cells can read it. It tells the cell how to make protein. This information is written in the DNA in a kind of code—a special language for describing proteins. This "genetic code" is arbitrary in the same sense that human language is arbitrary. For example, nothing forces English speakers to use the word "bird" to refer to animals with wings and feathers. It is merely a convention. Spanish speakers get the same idea across by saying "pajaro." Different species could have different genetic codes just as easily as different human populations speak different languages. But they don't. Every living thing—from the smallest microorganism to the largest whale—makes protein using essentially the same arbitrary code. What sense does this make, unless all these forms of life evolved from a single ancestor?

In writing this chapter, I find myself tempted to go on and on with evidence of common ancestry. It pervades all categories of biological data, yet I have mentioned only a few. But perhaps by now you get the point: Phylogenetic pattern is everywhere in nature. It makes sense only if all living things evolved from a single ancestor. And if life did have a single origin, then there can be no doubt that evolution makes large changes.

Conclusion

We began this chapter with a quote from Dewar, who couldn't imagine how whales could have evolved from land animals and concluded from this that they hadn't done so. This argument may once have seemed convincing. It collapsed, however, when paleontologists found the fossils of transitional species such as *Ambulocetus* and *Rodhocetus*. These animals were nothing like the creatures that Dewar imagined, and they apparently thrived along the ancient coast of Pakistan. A few years later, transposons dramatically confirmed the fossil evidence, showing that whales are indeed close relatives of sheep, deer, cattle, and especially hippopotamuses.

Dewar was wrong not only about whales, but also in a deeper sense. Because he could not imagine how whales had evolved, he concluded that they had not done so. This is the "argument from personal incredulity," which we met above on p. 18 in the context of Dewar's 1931 quote. Although that quote is 80 years old, the argument is no outdated antiquity. It is widely used even today in the literature of creationism. It has not improved with time. As the whale example shows, an argument based on incredulity is no argument at all.

4   CAN EVOLUTION
EXPLAIN DESIGN?

Every feather is a mechanical wonder. ...Whoever examines a feather, cannot help taking notice, that the threads or laminæ ... in their natural state unite; *that their union is something more than the mere apposition of loose surfaces; that they are not parted asunder without some degree of force; that nevertheless there is no glutinous cohesion between them, that, therefore, by some mechanical means or other, they catch or clasp among themselves, thereby giving to the beard or vane its closeness and compactness of texture. Nor is this all: when two laminæ, which have been separated by accident or force, are brought together again, they immediately* reclasp; *the connexion, whatever it was, is perfectly recovered, and the beard of the feather becomes as smooth and firm as if nothing had happened to it.*

William Paley, 1809 [86]

Imagine the ideal chemical refinery. It would convert abundant, renewable resources into a product that humans value. It would be smaller than a car, mobile so that it could search out its own inputs, capable of maintaining the temperature necessary for its reactions within narrow bounds, and able to automatically heal most system failures. It would build replicas of itself for use after it wears out, and it would do all of this with little human supervision. All we would have to do is get it to stay still periodically so that we could hook up some pipes and drain off the final product.

This refinery already exists. It is the milk cow.

Paul Romer, 2007 [96]

These quotes were separated by two centuries, but there is little difference in the sentiments they express. Both authors drew inspiration from nature, but not from her beauty. They were inspired instead by her engineering. Examples are not hard to find, for every living thing must solve many engineering problems just to stay alive and still more in order to reproduce. Look closely at any plant or animal, and you will find delicate engineering. The question is, where does it all come from?

Two hundred years ago, most people—scientists and laymen alike—would have agreed that the engineering came from God. This was the point of Paley's passage about feathers, which I quoted above. The anatomy of the feather, along with that of the eye and many other organs, showed evidence of design. For Paley, this implied a supernatural designer. It implied, in other words, the existence of God.

As a college student, Charles Darwin found this argument compelling. Years later, he still saw Paley as the single most useful part of his university education [54, p. 35]. But he no longer accepted Paley's argument, for by then he had conceived of an alternative: the evolutionary force of natural selection. In Darwin's view, natural selection is the mechanism by which plants and animals adapt to each other and to their environments. The resulting adaptations look to us like delicate engineering. .

This chapter will ask whether natural selection is a credible explanation of all this. It begins with simple forms of adaptation.

Simple adaptations

In January 1977, Peter Boag and Laurene Ratcliffe were by themselves on a desert island, surrounded by miles and miles of ocean. The island was tropical, but it was no paradise. A rocky, volcanic cinder cone, it was so small that you could walk all the way around it in an hour. There was no easy way onto the island, because the waves had carved its edges into sheer cliffs. Most years there is little rain, so the island had neither trees nor shade—just baking heat, black volcanic rock, and a few species of shrubby plants, insects, and birds. Boag and Ratcliffe were there for the birds.

The island is called Daphne Major (Figure 4.1) and is part of the Galapagos Islands of the eastern Pacific. Darwin visited these islands in 1835 and was interested in the birds there, including the various finch species now known collectively as "Darwin's finches." Daphne Major is home to several of these species, but our story concerns just one: the medium ground finch (*Geospiza fortis*), which is shown in Figure 4.2. Boag and Ratcliffe were part of a research

Figure 4.1: Daphne
Major in the
Galapagos. From
J. Weiner. *The Beak of
the Finch*. Vintage Books,
New York, 1994, by permis-
sion of K. Thalia Grant and
Jonathan Weiner.

team led by Peter and Rosemary Grant of McGill University. The team had studied these birds for years. Every year, they trapped finches, weighed and measured them, and attached colored rings to their legs. Each bird got a unique
combination of rings, so that the Grants and their students could identify individual birds by sight. By January 1977, they had trapped and measured most of
the finches on the island.

Boag and Ratcliffe were waiting for rain. Both of them had planned experiments for the breeding season. Boag wanted to measure the resemblance between parent and offspring, and Ratcliffe wanted to study the territorial calls of
breeding males. But neither experiment was possible untill the breeding season, and the birds do not breed until the rains come. All of them—birds and
scientists alike—waited for rain.

During 1977, the rains did not come. Boag and Ratcliffe recorded just 24 mil-
limeters during the breeding season, from January through May. Leaves shriv-
eled and dried, plants failed to set seed, and insect populations crashed. There
was almost nothing for a finch to eat. The birds scratched and pecked at the
shallow, rocky soil, searching for seeds left over from previous years. In each of
the five preceding years, the Grants and their students had sifted soil to count
seeds and measure their size and hardness. Boag and Ratcliffe continued this
chore during the drought. With each passing week, seeds became harder to
find. The most desirable seeds were small, soft, and easy to eat. As these disappeared, the finches turned increasingly to seeds that were larger and harder.

The larger and harder the seed, the longer it takes to crack the hull and
extract the nut meat. The smaller finches had most trouble, and they died in
droves. By the end of 1977, when the drought finally broke, the vast majority
of the medium ground finches on Daphne Major were dead. By Christmas,

35

Figure 4.2: Two examples of the medium ground finch (*Geospiza fortis*). They differ in the vertical dimension, or "depth" of their beaks [115]. From J. Weiner. *The Beak of the Finch*. Vintage Books, New York, 1994, by permission of K. Thalia Grant and Jonathan Weiner.

only 388—about one in seven—survived. These survivors were not a random sample of the original population. They tended to be large, with deep beaks, because such individuals were best able to eat the few remaining seeds. The average beak was deeper (measured from top to bottom) after the drought than before [10].

The Grants and their students had documented change, but they could not yet claim that it was evolutionary change. After all, the finches that they counted in December had also been there in January. Evolution is about change across generations, so the researchers needed to see another generation. When the surviving finches finally bred, Boag compared the beaks of the parents with those of their offspring. He found that parents with deep beaks tended to produce offspring with deep beaks [9]. Because of this resemblance between parent and offspring, the drought of 1977 produced real evolutionary change. The beaks became deeper not just in that first year, but in the generations that followed.

These changes did not however last forever. For the next few years, food was plentiful, and there was no advantage in a large body or a deep beak. Selection began slowly to restore these characters to their original values. By 1985, the finches looked much like their ancestors had before the drought. The island was hit again by drought in 2004, and this one was even more devastating. By year's end, only 83 medium ground finches remained alive. But this time, the result was very different: the survivors had smaller beaks rather than larger ones [50]. Once again, the change made adaptive sense. By 2004, the island had been colonized by the large ground finch (*Geospiza magnirostris*), which monopolized the large, hard seeds. For the medium ground finch, the only hope lay in the very smallest seeds. Those that survived had smaller beaks, which enabled them to find small seeds in crevices among the rocks.

Over the years, the beaks of these finches have been shaped and re-shaped by the process that Darwin described over 100 years before:

> Of all birds annually born, some will have a beak a shade longer, & some a shade shorter, & that under conditions or habitats of life favouring longer beak, all the individuals, with beaks a little longer would be more apt to survive than those with beaks shorter than average. [51, p. 46]

In the present context, the important point is that these changes were improvements. In 1977, it was good to have a deep beak, so selection made beaks deeper. In 2004, it was better to have a small one, so selection made beaks smaller. Evolutionists think this is how all adaptations arise.

The finches on Daphne Major illustrate Darwin's argument in action. For many years, this argument was only a hypothesis. In 1911, there was good evidence that evolution had in fact happened. There was far less evidence, however, that natural selection was an important mechanism. As one skeptic put it: "A conclusion may be perfectly logical and still not true" [87, p. 117]. Such criticism is no longer reasonable. We can watch every step of the process, and we can see in detail how it produces adaptation. These facts are no more in doubt than the theory that the earth is round.

Yet there is still room for skepticism. Observations of selection in action tend to involve simple changes—change, for example, in the depth of a bird's beak. This is understandable, because such changes are faster and thus easier to observe. But could it be that these simple changes are the only ones that natural selection can really make? It is easy to see how a blind process might create a small change in the shape of a beak, but harder to imagine it engineering anything really complex. Let us take a close look at a complex adaptation.

A complex adaptation: the eye

Your eye is busy as you read this page. At a minimum, it must turn light energy into nerve impulses. This is the job of the retina, which lies at the back of the eye. But you still could not read without focusing the light into an image. Your eye does this with the cornea and lens, working together. This is not simple, because the rays of light must be bent differently as your gaze shifts from near to far. Just now you are focusing *right here*, on words perhaps 15 inches from your nose. To accomplish this, small muscles distort your lens into just the right shape. This gets an image onto the retina, but you still could not read if that image were too bright or dim. Thus, the muscles in the iris adjust the pupil so that it lets in just enough light. Yet another set of muscles points your eye at the page. If any of these components had failed, you would not be reading.

How could all this have evolved by natural selection? In 1859, Darwin spent several pages on eyes. He later confessed to a friend that the subject still gave him a "cold shudder." He was right to shudder, for the eye has always been central to the argument against his theory. The argument about eyes and evolution begins with an unlikely character: the man who taught science and mathematics to Darwin's sons. Charles Pritchard was headmaster of Clapham Grammar School. It must have come as a shock when the *Origin* appeared, and the shy, retiring father of his pupils George and Francis Darwin became a sudden celebrity. Pritchard was a man with two passions—religion and astronomy—which he viewed as a single passion. Indeed, all of science and religion was for Pritchard a seamlessly unified whole. There was not, nor could there ever be, any possibility of conflict between the two. Yet conflict there was as soon as Darwin's new theory appeared in print.

Pritchard got a chance to set the record straight in 1866, when he preached a sermon before the British Association for the Advancement of Science. In an appendix to the published version of that sermon, Pritchard argued that the vertebrate eye could not have evolved by natural selection [90]. As he emphasized, the various parts of the eye are delicately adjusted to each other. Change one part without changing the others, and this balance is ruined. The eye will no longer see as well. To improve an eye, Pritchard argued, one must change all the parts at once. Furthermore, these changes must be delicately coordinated, each part changing by just enough to complement the changes of the others. For mutation, acting blindly, to produce such coordinated changes was in Pritchard's view

> not less improbable, than if all the letters in the *Origin of Species* were placed in a box, and on being shaken and poured out millions on millions of times, they should at last come out together in the order in which they occur in that fascinating and, in general, highly philosophical work. [90, p. 33]

Pritchard preached his sermon in August and did not have to wait long for the printed version. He sent a copy to Darwin in early October, and he must have circulated it widely. It seems likely that one copy made its way to Joseph John Murphy in Belfast. Murphy had just been elected president of the Belfast Natural History and Philosophical Society. In mid-November, he delivered his presidential address to that society [74]. For the first 20 minutes, Murphy sounded like a Darwinian. He piled fact upon fact, arguing that the evidence in favor of evolution was overwhelming. But no one would have mistaken him for a Darwinian in the second half. Murphy accepted natural selection as a fact, but he doubted that it was very important. He saw no way that natural selection could generate the complex engineering present in every organism. Like Pritchard, he focused on the vertebrate eye. He even paraphrased Pritchard's argument. But he also introduced a new one, which has turned out to be more important. After summarizing the anatomy of the vertebrate eye, Murphy said

> Here is a considerable number of parts and connections, every one of which would be useless without all the rest, and every one of which, consequently, pre-supposes all the rest. [74]

If the eye evolved gradually, then early eyes must have been incomplete: perhaps a retina with no lens or a lens with no retina. These incipient eyes would, in Murphy's view, have been useless. They could not have been shaped by natural selection into the complex eyes with which you are reading this page. This argument has come down to us in the modern creationist literature that asks "What good is half an eye?" If it holds for eyes, then it surely holds for any complex character. Had Murphy been right, it is hard to see how natural selection could have engineered any complex organ, let alone a whole organism. Yet Murphy should have known better. For as we shall see, Darwin had undermined this argument (and Pritchard's too) seven years before, in the first edition of *Origin*.

Is it *plausible* that a complex eye could evolve?

The arguments of Murphy and Pritchard were both about plausibility. It is *implausible*, they claimed, that natural selection could produce a complex eye. The nice thing about such arguments is that one can attack them without any data at all, merely by inventing a plausible story about how complex eyes *might* evolve. The story only needs to be plausible; it does not need to be true. Thus, the first piece of Darwin's argument was a made-up story about how eyes might have evolved.

I often tell my own version to introductory classes. It begins with a hypothetical creature that lives in water and eats food that grows in light. It must swim toward light to find food. It accomplishes this using a simple "eye spot," which is nothing more than a cluster of light-sensitive cells. This eye is illustrated in panel A of Figure 4.3. Using this eye, the creature forages using a simple rule: when light strikes the eye spot, swim forward; otherwise turn.

The eye spot helps but has limitations. It cannot tell whether a light source is straight ahead or off to the side. The creature will swim straight ahead and may miss the bright patch altogether. It would get more to eat if its eye spot were sensitive only to light from straight ahead. This problem is easy to fix: put the eye spot into a shallow cup as shown in panel B. The eye is now shielded from the side and responds only to light coming from more or less straight ahead. The creature finds food more easily, because it is more likely to swim directly toward bright patches.

This new eye is better but still far from perfect. It can tell that the light source is somewhere up ahead, but that is all it can tell. It cannot tell whether the light is a bit to the left or a bit to the right. We can improve its directional capability by deepening the eye pit, as shown in panel C1. The eye is now so deep that it is sensitive only to light from straight ahead. It has become a pinhole camera, forming an image on the eye spot (which we can now call a retina). Not only can this creature sense light, it can really see.

Yet sadly, not very well. In solving one problem, the new design has caused another. It lets in so little light that this creature perceives its world as dim. Perhaps this is why there are so few pinhole-camera eyes in nature [80]. To improve things, we need somehow to gather more light without sacrificing directional ability.

This brings us to panel C2. Rather than deepening the eye cup, this eye has secreted a glob of transparent mucus into it. The mucus is a little denser than water, so the light rays passing through it bend inward toward the center, and the mucus serves as a primitive lens. This lens does not have to form an image or even bend the light rays much. Bend them just a little, and you have increased the animal's direction-finding ability without making its world dim.

From here on, the evolutionary path is easy to see. The animal benefits from every increase in quality of lens or retina. Eventually, the improved lens will focus light into a sharp image. The result is a complex camera-type eye, which evolved from a simple eye spot by a succession of small, individually adaptive changes.

This story is of course a fabrication. The cartoons in Figure 4.3 are just that—cartoons. By what right do I claim the story is plausible? Could creatures with eyes like those ever really live? The answer to this question is an unambiguous "yes," for all these sorts of eye can be found in creatures alive today. This was the second piece of Darwin's argument, and the evidence is even better today. Figure 4.4 shows the eyes of several snails. You will recognize cup eyes and

Figure 4.3: Hypothetical steps in eye evolution. Artwork copyright © (2011) by Paul Kidby, www.paulkidby.net.

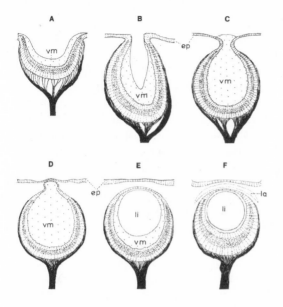

Figure 4.4: Eyes of snails and slugs. A, eye pit; B, cup eye; C, pinhole-camera eye; D, closed eye; E, F, eyes with lenses. From Fig. 11, p. 237 of L. von Salvini-Plawen and E. Mayr. On the evolution of photoreceptors and eyes. In M.K. Hecht, et al., eds., *Evol. Biol.*, 10:207–263. Plenum Press, New York, 1977, with kind permission of Springer Science and Business Media.

pinhole camera eyes. In several eyes there is a stippled layer labeled *vm*. This stands for "vitreous mass," the technical term for the gooey mucus I described above. These snails do not represent a sequence of ancestors and descendants. (How could they? They're all alive today.) Yet they leave no doubt that all these eyes can work in real animals.

My version of Darwin's story begins with a very simple eye. Yet one might object that there is a lot of complexity even here. Doesn't an eye require a so-phisticated brain and a complex of nerves that connects it to the eye? Apparently not. Box jellyfish have eyes but no brain. Their eyes are directly connected via nerves to the muscles of their tentacles. And vision of a sort can be found in even simpler creatures. *Chlamydomonas* is a genus of single-celled green algae. With the help of an eye spot and two flagella, these tiny creatures swim toward light [42].

One might also worry about time. Could the changes in my fabricated story really happen fast enough? This problem is worse than it seems, for if eyes evolved at all, they did so many times. This is another argument that goes back to Murphy, who pointed out that complex eyes are found not only among vertebrates, but also among worms and mollusks. The most primitive vertebrates have only rudimentary eyes, while the most primitive worms and mollusks have no eyes at all. Murphy argued that complex eyes, if they evolved at all, must have evolved at least three times [75, p. 321]. Modern comparative anatomists agree, but extend the number to at least 40 and probably more like 60 [112]. If eyes evolve, they must do so often and easily. Could it really be so easy?

Dan-Eric Nilsson and Susanne Pelger have answered this question. They constructed an evolutionary story much like the one that I told above. Theirs was more detailed and was constructed so that each change increased the "spatial resolution" of the eye. (Think of this as the number of pixels in the image.) The eye in their model is just a list of numbers. One describes the depth of the eye pit, another the refractive index of the central portion of the lens, and so on. Initially, the eye is simply a patch of light-sensitive cells covered by a clear layer of protective cells. Each generation, selection changes one of these numbers by a small amount (1/200 of a percent). The number that changes is the one with the largest effect on spatial resolution. In their model, it took 363,992 generations to evolve from a simple eye spot to a complex, camera-type eye with optimal optics. In the real world, these evolutionary changes occurred in small aquatic creatures with generation times of a year or less. Thus, the whole process probably took less than half a million years [79]. It no longer seems so surprising that complex eyes have evolved so often.

So far I have argued more or less as Darwin did, and I hope I have convinced you that eyes might plausibly evolve. This is enough to demolish the arguments of Pritchard and Murphy, but it does not tell us whether eyes really did evolve. We cannot answer that question by making up stories. We need real evidence—evidence that Darwin did not have. If eyes did evolve, then closely related species should have similar eyes. Their eyes, in other words, should show traces of common descent. These traces may show up in the morphology of eyes, in the proteins of which they are built, or in the underlying DNA. These different approaches tell very different (but complementary) stories.

Traces of common descent stare back at you whenever you visit the zoo. The lion, the ostrich, the iguana, the bullfrog, and the goldfish all have eyes a lot like yours. Other animals (as we shall soon see) have eyes as strange as anything in science fiction. These anatomical differences are easy to see, but their evolutionary history is complex. Let us start instead at the other end of the subject: with the molecules that make vision possible.

Traces of common descent in opsins

To see, you must have cells that sense light. Nature might have made such cells in a variety of ways. For example, many organisms—plants as well as animals—respond to sunlight by manufacturing proteins that repair DNA. This is a good idea because the ultraviolet component of sunlight damages DNA.

This chemistry has also been modified for other purposes, which range from helping plants turn toward the sun to helping animals go to sleep at night and wake up in the morning [15]. Nature might have used this same chemistry to make eyes, but it never has. Instead, every animal that sees does so with a single family of proteins called opsins. Why? A creationist might argue that opsins are simply the best way to do the job and were thus selected for this job by the creator. An evolutionist, on the other hand, would see these similarities as further evidence that all animals had a common ancestor, which must have responded in some fashion to light.

How can we tell which account is correct? If the evolutionists are correct, then we should also see traces of common descent at finer scales. Closely related species should have closely similar opsins, and distant relatives should have opsins that are less similar, just as we saw with other characters in Chapter 3. These patterns do exist, but they are overlain by another sort of historical record. To appreciate the history of opsins, you must first understand why there is more than just one kind of opsin protein.

The machinery of reproduction is like a Xerox machine for duplicating chromosomes. This machinery often stutters, and some stretch of a chromosome gets copied twice. If that stretch happens to contain a gene, then the new chromosome will carry a double dose of that particular gene. In this way, chromosomes with duplicated genes are introduced into populations. Over time, some duplicates evolve new functions and take on lives of their own. When this happens repeatedly, we end up with a "family" of related genes, all of which evolved by gene duplication from a single ancestral gene. The opsin genes are one such family.

Your eyes contain four kinds of opsin—four members of the opsin family of genes. One of them is very sensitive and allows you to see in dim light. The other three respond differently to light of different colors. By comparing the output of the photoreceptors that contain these three opsins, you are able to see in color. This system of three-way color vision is something that you share with Old World monkeys and apes but not with New World monkeys or other mammals. These have just two kinds of color receptor and cannot distinguish as many colors. This is one indication that we share a common heritage with Old World monkeys and apes.

If this is really true, then one of our opsin genes must have arisen fairly recently—after the split with New World primates—by gene duplication. All evidence suggests this is so. Two of our opsin genes, the ones most sensitive to long and intermediate wavelengths, are located close together on the X chromosome. Their protein sequences are very similar to each other and also to the long-wavelength opsins of other mammals.

The howler monkey (*Alouetta sp.*) is an exception to all this. It is a New World monkey but has three kinds of color receptor like an Old World monkey. Furthermore, two of these lie on the X chromosome and specialize in light of medium and long wavelengths. In all these respects, the howler is like an Old World monkey. But it is *not* an Old World monkey. It has the prehensile tail, the flat nose, the sidewards facing nostrils, and even the DNA of a New World monkey. So why does it have the color vision of an Old World monkey?

This mystery melts away when you examine the DNA in detail, especially the regions just upstream and downstream of the duplicated gene. These flanking regions were also copied when the gene was duplicated. In the howler, these regions are like the corresponding regions in other New World monkeys; they are quite different from those of Old World primates. The flanking regions also allow us to identify the boundaries of the chromosome fragment that was inserted when the gene was duplicated. In the howler, these boundaries are unlike those in Old World monkeys. It appears that this opsin gene duplicated twice. One duplication happened in the ancestor of Old World monkeys and apes, and the other happened in the ancestor of howlers. The howler duplication was apparently more recent, for the two opsin genes on its X chromosome are nearly identical [105].

Our system of three-way color vision is unusual among mammals. Most mammals have just two kinds of color receptor and are, to our way of thinking, color blind. In this respect, mammals are unusual among vertebrates, most of which have four kinds of color receptor. To a lizard, a bird, a frog, or a fish, we are the ones who are color blind.

In all of these patterns, opsins (and their associated color receptors) show the traces of common ancestry. There are similarities among Old World monkeys and apes not shared by other mammals, similarities among mammals not shared by other vertebrates, similarities among vertebrates not shared by invertebrates, and so on. This nested pattern of similarities goes all the way back to original opsin, with which this story began. If these data had been available to Darwin, he would have had no reason to shudder about eyes.

Traces of common descent in developmental genes

There are also traces of common descent in genes that control the development of eyes. Genes are the blueprints from which proteins are made. Some encode proteins used as building blocks—the stuff of which your body is made. Many however act as switches, turning other genes on and off. This is what makes one part of your body different from another. Brain cells differ from muscle cells because they express different genes, and the same goes for eyes. Your eyes became eyes because a suite of these molecular switches silenced some genes and directed others to make protein. These are called "regulatory genes," and we know a lot about several of those involved in making eyes. At least three of these, *pax6*, *six*, and *otx*, direct the construction of eyes in animals as different as mammals and insects (see also p. 30).

You may be wondering how we know that the versions of these genes in such different animals are really the same. Part of the answer lies in the similarity of the proteins they encode. The *pax6* proteins of mouse and human are identical, and those of mouse and fruit fly are very similar. But this is only part of the story. A few years ago, Walter Gehring and his colleagues found a way to insert the mouse version of *pax6* into a fruit fly embryo. This tricked the fly embryo into expressing the mouse gene. The gene was not expressed throughout the fly but only in the tissues that Gehring's group specified. In different experiments, they specified different tissues. They got flies with eyes on their wings, eyes on their legs, and eyes on their antennae. These were not mouse eyes, even though they had been induced by a mouse gene. They were the usual compound eyes of an insect. The mouse gene had simply switched on the fly's machinery for making eyes. Not only does this work with the *pax6* gene of a mouse, it even works with that of a sea squirt.

There are major differences among sea squirts, squid, fruit flies, and mice. Their eyes could hardly be more different. Yet in spite of their differences, all these eyes are built around a similar chemistry for detecting light, and all are constructed in response to the same regulatory genes. It is hard to avoid the conclusion that these disparate eyes evolved from some common ancestor that could respond to light. Its visual system was probably no more complex than the simple eye spot in panel A of Figure 4.3. But it was surely assembled using *pax6*, and it surely used opsins to detect light.

Traces of common descent in lens proteins

We are just getting started, for eyes are not built only of opsins, and every protein has its own historical story to tell. Lenses are built of transparent proteins called *crystallins*, and there are traces of common descent here too. The same crystallins, for example, are found in all vertebrates. Yet this pattern does not reach all the way back to the common ancestor of people and squid. Our eyes look a lot like those of squid, but our crystallins are unrelated. Lenses must have evolved *after* the common ancestor of vertebrates and squid. That common ancestor must have had some sort of eye, but probably not one with a lens.

Traces of common descent in eye morphology

If this common ancestor lacked a lens, then its eye must have been very simple. It probably didn't have an iris or cornea or any of the other anatomy that we see in the eye of another human, or for that matter in that of an insect. If lens proteins evolved late, then this visible anatomy must have evolved late too. We should find traces of common descent only among closely related animals. Let us test this hypothesis by examining different kinds of animal.

Vertebrates

Most of the animals in a typical zoo are vertebrates and, as I mentioned above, have eyes a lot like ours. Our eyes, like those of fish, frogs, lizards, birds, and other mammals, are miniature cameras. With one exception (which we'll get to in a moment), you don't find this type of eye elsewhere in nature.

Crustaceans and insects

The *arthropods* include crustaceans (lobsters and their relatives) and insects. These animals resemble each other not only in having exoskeletons and jointed appendages, but also in the structure of their eyes. Take for example the gall midge shown in Figure 4.5. Its eyes consist of many tiny units, each with its own lens and photoreceptors, and each pointing in a slightly different direction. The midge's brain constructs an image by integrating the output of all these tiny eyelets. This sort of eye is common to all arthropods but is not found in other animals.

Snails

Then there are the snails. Figure 4.4 shows a few of the kinds of eye found among snails, and no one could argue that these are all the same. They range from simple eye pits to complex eyes with lenses. One group, the heteropod sea snails, has retinas so long and narrow that at any given instant the snail's field of view is only a few degrees high, though it may be 180° wide. This snail peers at the world through a narrow slit. The slit however is constantly in motion. It tilts up and down, covering 90° in about a second. This allows the snail to detect prey within a window roughly 90° high and 180° wide [67].

Why should snails have such a diversity of eyes? It is not that traces of common descent are wholly lacking, for closely related snails have similar eyes. Heteropod sea snails, for example, all have long narrow eyes. Yet if we knew only about the anatomy of their eyes, no one would think all snails were relatives. On the other hand, every other category of evidence argues that snails are related.

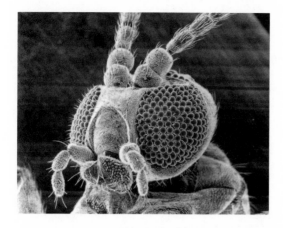

Figure 4.5: Compound eye of gall midge, a tiny insect whose larvae cause galls in plants. The head is about 0.3 mm (0.01 in) wide. From http://remf.dartmouth. edu/images/insectPart3SEM/ source/14.html.

We all know a snail when we see one. Their relatedness is manifest not only in these obvious features, but right down to their DNA. It seems clear that snails evolved from a single ancestor. Why then are their eyes so different?

The obvious answer is that the ancestral snail lacked eyes. Molecular evidence suggests that this ancestral snail could detect light, but probably not much more. It had nothing as sophisticated as the organs we call "eyes." Consequently, the various kinds of eye in modern snails cannot be traced back to a single ancestral eye. They are independent inventions. Is this evidence consistent with that of the lens proteins? We don't know, because these proteins haven't been described. We can only wait and see.

Cephalopods

Squid and most other cephalopods present us with a different sort of problem. If you looked only at its eye, you might think that a squid was a vertebrate, for its eye like ours is a little camera. Yet the relationship between squid and vertebrate is remote. It is like the three-way color vision of the howler monkey, which I discussed above. Both cases show a close similarity between two unrelated animals. Yet in both cases, this similarity melts away when you examine it closely. Not only are the eyes of cephalopods and vertebrates made of very different molecules, they also differ in anatomical detail. The photoreceptors of a squid, for example, face toward the light. Ours face away. The eyes of squids and vertebrates are only superficially similar. Natural selection seems to have invented the camera-type eye more than once.

In summary, the pattern for visible morphology is like that for lens proteins. Major groups of animals have similar eyes, yet these similarities do not extend to comparisons *between* major groups. This pattern is known in exquisite detail and is replete with evidence of common descent.

Creationists sometimes argue that species are similar not because of common ancestry, but because of design constraints. For example, the creationist argument that I proposed on p. 43 suggested that all animals use opsins to detect light because they are optimal for that job and were thus selected for it by the creator. In the same way, one might argue that all vertebrates have similar eyes because that is the best way to make an eye. This however would ignore the very different eyes found in snails, cephalopods, mollusks, and insects. Because so many types of eye exist, one cannot argue that there is only one way to make one.

Furthermore, many of the eyes found in nature are demonstrably suboptimal [84]. For example, the network of arteries and veins that supplies our own retinas lies in front of the retina and obstructs vision. There is no necessity in this, for other animals have the blood supply behind the retina and out of the way. In spite of this deficit, our own eyes (and those of other vertebrates) are a lot better than those of many other animals. The *Nautilus* makes do with the pinhole-camera eye described on p. 40. These suboptimal eyes make it impossible to argue that the creator has selected for each species the type of eye that works best.

On the other hand, evolution explains the entire pattern. Some parts of the eye—opsins and developmental genes—evolved early and are thus shared by animals as different as vertebrates, squid, snails, and insects. Others parts— lens proteins and eye morphology—evolved later and are thus shared within these categories of animal but not between them. Not only did eyes evolve, they evolved more or less as proposed in Darwin's "just so" story, which I outlined above on pp. 39–40. Evolution of complex eyes is not merely plausible; it is also true.

Irreducible complexity

We began this section with Charles Pritchard and Joseph John Murphy, who argued that natural selection was incapable of producing any complex organ with interacting parts. In particular, they argued that it could never have produced the vertebrate eye. If this were true, it would demolish the case for evolution by natural selection. But it is clearly false. The arguments of Pritchard and Murphy may have seemed plausible in 1866, but not any more.

Nonetheless, their argument is still in circulation, and Michael Behe has given it a new name: *irreducible complexity* [7]. For Behe, an organ is irreducibly complex if it involves several parts, all of them essential to its function. Natural selection, he claims, could never assemble such an organ. Early versions would be incomplete and therefore could not function. Natural selection would weed them out.

I hope this argument strikes you as familiar, for it is precisely the one that Murphy made in 1866 (see p. 39). Murphy saw no way that one part of the eye could function without the others, but it turned out that he was wrong. Retinas for example can function without lenses, and we understand in considerable detail how eyes were assembled by natural selection.

Behe's claims have suffered a similar fate. Like Murphy, he identified adaptations that he thought natural selection could not build [7]. For example, many hormones fit together in pairs like a lock and key. What good is the lock without the key? How can one evolve before the other? Jamie Bridgham and his colleagues studied one such pair and found that the key evolved first—it formerly interacted with a different molecule. They even worked out the precise mutations that gave rise to the current lock-and-key interaction [1].

Behe's most famous argument involves the bacterial flagellum, a structure like a tiny tail that helps the bacterium move. It is a complex affair, involving 30 proteins. According to Behe, all are essential and none has any other function. A partial flagellum would thus serve no purpose and could not evolve by natural selection. It is Murphy's argument all over again. But this argument has also collapsed. It turns out that a simpler organ, involving several of the same proteins, is used by some bacteria to inject toxins into other cells [70]. It is simply not true that these proteins have no function apart from their role in the flagellum. Behe was wrong about the flagellum just as Murphy had been wrong 140 years earlier. His argument is another example of Dawkins's "argument from personal incredulity" [31, p. 38]. That argument is even weaker now than in 1866.

Conclusion

Natural selection is easy to understand, but it is remarkable that a process so simple can produce feats of complex engineering. It is easy to see how people might be skeptical. The basis for this skepticism was laid out in the 1860s and has not changed since: If evolution is gradual, then intermediate forms must be incomplete, and incomplete forms must be maladaptive. Because of these maladaptive intermediates, complex adaptations cannot evolve by natural selection.

Yet all evidence supports the view that natural selection *is* responsible for nature's delicate engineering. The arguments of Pritchard, Murphy, and Behe all fail in the same way—by taking a rigid view of function. Murphy saw no value in an eye that failed to form a well-focused image. Yet some eyes form no image at all; they merely sense light. Behe saw no value in flagellar proteins apart from their role in the flagellum. Yet some bacteria use them to inject toxins. Murphy and Behe failed to imagine these alternative functions. Natural selection cobbles adaptations together using the materials at hand. For example, it built the lenses in vertebrate eyes from proteins that evolved originally as enzymes and are still used that way elsewhere in the body. Over time, proteins may acquire new functions and lose old ones. There is no reason to assume that a protein used today in eyes or in a bacterial flagellum evolved originally for that purpose.

The example of the eye proves that even a complex organ can evolve by intermediate steps that are all adaptive. But is this the *only* way? Do they ever pass through intermediate stages that really *are* maladaptive? Many evolutionists think so, as we'll see in the next chapter.

5

PEAKS AND VALLEYS

C onsider this sequence of words: "rat," "rut," "rum," and "sum." Each word differs by one letter from the word before it, and each is correctly spelled and means something. Yet the last word differs from the first at every position. If you think this game is easy to play, try beginning with a longer phrase, such as "brain of a rat." It no longer works to change "rat" into "rut," because "brain of a rut" doesn't mean anything. We could change "rat" into "bat," but there are not many such possibilities. There is no way I can find to produce a new phrase that differs at every position. This difficulty reflects the fact that we began with a longer phrase. The longer the word or phrase, the harder it is to transform.

Michael Denton suggests that this, in microcosm, is the problem of evolving complex adaptations [34, Chapter 4]. The more complex the adaptation, the less likely it is to be useful in an incomplete form. And unless it is useful, it cannot spread by natural selection. There are, of course, complex adaptations within every species, so Denton argues that each species is trapped on an "island of function." Evolution can move a species around on its island but cannot move it to another island. From time to time, mutant individuals that lie between islands may arise. Such individuals would have a mixture of features that work poorly together. They would therefore be eliminated by natural selection. For this reason, Denton argues that selection can make minor adjustments to existing adaptations but cannot create complex adaptations from scratch.

This argument hinges on the idea (introduced in Chapter 3) that intermediate forms may be maladaptive. This idea has a long history, going back to Georges Cuvier [24, p. 11], who opposed evolution decades before Darwin, and extending to authors such as Michael Behe [7], who oppose evolution today.

Evolutionists have responded to such arguments in several ways. One response points to complex adaptations such as the vertebrate eye. As discussed

51

in Chapter 4, the eye seems to have evolved by a series of small changes, each of which was advantageous. This example refutes Denton's claim that complex adaptations *cannot* evolve without passing through maladaptive intermediate stages.

It is tempting to stop here. Denton's central claim is false, so what more is there to say? In fact, there is rather a lot. Many evolutionists would *agree* with Denton's central point—that the evolution of complex adaptations often requires passing through maladaptive intermediate stages. It seems dishonest to pretend that the whole idea is merely silly. Denton has raised a good question (several really), so this chapter will explore what evolutionists have had to say.

Like Denton, evolutionists are concerned with the relationship between the characteristics of an organism and its fitness. To be concrete, let us talk about the length and depth of beaks within some hypothetical species of bird. Presumably, some combinations of length and depth are better than others, and this will affect the birds' fitnesses. In this context, a length/depth combination with high fitness is one that allows birds to leave lots of descendants.

Imagine making a graph, with beak length on the horizontal axis and beak depth on the vertical axis. Each combination of length and depth corresponds to a point on the graph. To represent fitness, we might draw contour lines as on a topographic map. Alternatively, we can use a perspective drawing, as in Figure 5.1. Either way, the graph represents what is called a *fitness surface*. The horizontal position on the graph refers to beak dimensions, and the vertical position to fitness. Combinations of beak length and depth with high fitness will graph as high points on the fitness surface.

The surface in Figure 5.1 has two peaks, one for beaks that are short and deep, the other for those that are long and shallow. Recall that among medium ground finches, short deep beaks were best for cracking hard seeds and long, shallow ones for probing in cracks. Something of this sort presumably accounts for the two peaks in our hypothetical surface.

So far, I have talked about this surface as though it described the fitnesses of individual birds, and that is certainly one way to think of it. But there is also another way, which is often more useful. In this second interpretation, each point on the surface refers to an entire population. The horizontal axes refer not to the beak dimensions of individual birds, but to the mean values of these dimensions. For example, a population in which the average bird has a short deep beak would plot near the lower of the two peaks in the figure. One in which the average beak was long and shallow would plot near the higher peak.

This second interpretation is useful because it gives us a way to think about evolutionary change. Natural selection always pushes populations uphill on

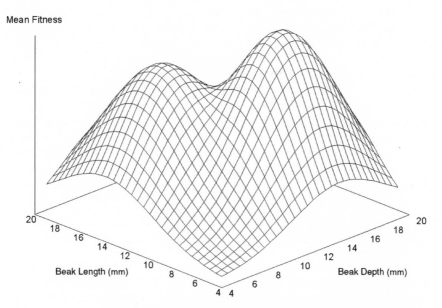

Mean Fitness

Beak Length (mm)

Beak Depth (mm)

Figure 5.1: Hypothetical fitness surface.

the fitness surface.[1] In other words, it favors the character values that increase mean fitness. The finches of Daphne Major provide a good example. During the drought, selection reshaped their beaks. The result was a population with members that were better able to eat the remaining food and thus had higher Darwinian fitness. We picture this as an uphill movement on the fitness surface that prevailed during the drought. This uphill trajectory is both the blessing and the curse of natural selection.

It is a blessing because it is the engine of adaptive evolution. Without it, evolution would not improve the fit between organisms and environments. It could never have produced anything as complex as a bacterium, let alone a mammal. But it is also a curse. To see why, imagine that a population starts somewhere on the left half of Figure 5.1 and then climbs uphill. It is likely to reach the peak on the left—the lower of the two. Once it reaches that peak, it

---

[1]Although selection always pushes straight uphill, this may not be the direction in which the population moves. For one thing, the population is jostled about by random forces. The smaller the population, the larger are these random shocks. Even in large populations, the movement may not be straight uphill because populations are easier to push in some directions than others. Just as a wagon is easiest to push in the direction that the wheels roll, populations are easiest to push in the direction with most genetic variance [4]. Nonetheless, they still move generally uphill.

will stop. It cannot then reach the higher peak without crossing a valley. But this would involve going downhill *against* the force of natural selection. So the population will stay there, stuck on the lower fitness peak, for a very long time. This is the curse of adaptive evolution.

The lack of foresight here is spectacular. Populations get stuck on inferior fitness peaks even when a superior peak is (so to speak) in plain view. This is not some defect in the theory of evolution. It is just how evolution works. Natural selection is an engine that improves adaptation, but it does not promise perfection. If populations evolve on a rugged fitness surface—one with lots of peaks and valleys—then adaptations should often be imperfect. But how do we know that the fitness surface *is* rugged? It is time to look at evidence.

Poor engineering

As we have seen, natural selection is a pretty stupid engineer. It simply walks uphill on the fitness surface. If that surface has lots of small peaks, selection is likely to get stuck on one. This will look to us like poor engineering. On the other hand, if the fitness surface were relatively smooth, poor engineering should be hard to find. This does not seem to be the case, however, for nature provides many examples of bad engineering.

Consider the gardener in Figure 5.2, who has run out of hose. For natural selection, this is an old story. As our ancestors evolved from fish into amphibians, reptiles, and mammals, their bodies changed enormously in size and shape. This required corresponding changes in the various tubes and wires—arteries, veins, nerves, and so on—that run throughout our bodies. In many cases these tubes became stretched around some obstacle, confronting selection with a dilemma like that of the gardener in the figure. All too often, it failed to do the sensible thing. Rather than walking back around the tree, selection got another length of hose.

One such example involves the vas deferens, the tube that carries sperm from testicle to penis [123]. In cold-blooded animals such as fish, the testes lie deep within the body cavity, not far from the heart. (Not a bad place, when you think of it, for an organ so delicate and so crucial to its owner's reproductive success.) Each testicle is a factory for making sperm. The process takes about five weeks (in humans) and is surprisingly sensitive to temperature. It works well in the body cavity of a shark but would work poorly at the higher temperatures that prevail inside our own bodies. As the bodies of our ancestors got warmer, it must have gotten harder and harder to make sperm. Selection would have favored individuals with testes a little closer to the skin, where things

Figure 5.2: The gardener has run out of hose. What would you do: (A) go back around the tree, or (B) add another length of hose? Natural selection chose option B. Copyright © 1996, 1998, George C. Williams. Reprinted from G.C. Williams, *Plan and Purpose in Nature*, by permission of Weidenfeld & Nicolson, an imprint of The Orion Publishing Group, London, and from reference by permission of Basic Books, a member of the Perseus Books Group [124].

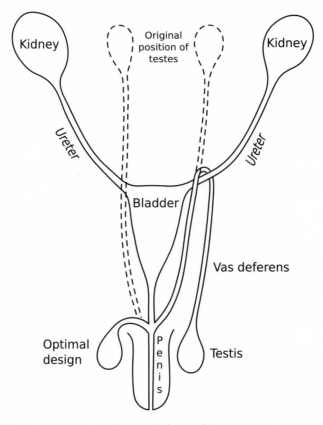

Figure 5.3: In human males, the vas deferens follows a circuitous route around the ureter, as shown on the right, instead of the direct route shown on the left [123, p. 142]. Redrawn from reference [123, p. 365].

were a little cooler. Over time, the testes migrated out of the body cavity and into the little pouch—the scrotum—that hangs between the legs of every male mammal. Along the way, however, the testes took a false turn. This took them the wrong way past the ureter, the tube that carries urine from kidney to bladder. The vas deferens got hooked there, just as the gardener's water hose is hooked around the tree. But rather than unhook the vas deferens, selection increased its length. The result is shown on the right side of Figure 5.3. It works, but it is not something an engineer would be proud of.

Other waterhose problems can be found among the nerves of the head and neck. In fish, the layout is sensible. Each gill arch is supplied by its own artery and its own nerve. These arteries and nerves all branch from a larger artery and a larger nerve, which run lengthwise down the fish. There is nothing in this arrangement to make an engineer wince.

These nerves and arteries can be traced back into the fish embryo, where the four gill arches are just wrinkles (called *pharyngeal arches*) at the base of the blob that will become the fish's head. At that stage, most vertebrate embryos have similar pharyngeal arches. The resemblance is more than superficial, for the arches develop into similar structures in adult animals. For example, your first arch develops into a jaw, whether you are a shark or a human. Even where mammalian anatomy is vastly modified, it is sometimes possible to trace the changes in the fossil record. For example, there is a clear sequence of fossils connecting the rear-most bones of a reptile's jaw to two tiny bones (the malleus and incus) in the human ear [100].

The structures that develop from each arch are fed by nerves and arteries that originate in that arch. In the nerves of a fish, this leads to a simple wiring diagram. In ours, it leads to a tangle that generations of anatomy students have learned to dread. The best example is the recurrent laryngeal nerve, which enervates the voice box and thus enables us to speak. It starts in the head, travels down to the chest where it loops around an artery, and then travels back up to the throat. In humans, it ends up only a few inches from where it started. This is another waterhose story. In fish, the nerve and artery both feed the rear-most gill arch. Over time, the artery descended into the chest, and the nerve went with it [33]. In giraffes, the nerve is 20 feet long, yet the direct route is only a foot. Engineers get fired for this sort of thing.

One can also find poor engineering in the retina of the vertebrate eye, which appears to have been installed backwards. The light-sensitive portion of the retina faces away from the light, but this in itself does not appear to be a problem [65]. It is unfortunate, however, that the nerves, arteries, and blood vessels that serve each photocell are attached at the front rather than the back. They run across the surface of the retina, obscuring the view. To provide a route for this wiring, nature has poked a hole in the retina, which causes a substantial blind spot in each eye. You don't notice these because your brain patches the image up, but that fix is only cosmetic. You still can't see any object in the blind spot, even if it is an incoming rock. Retinas don't *have* to be attached this way: those of cephalopods are attached more sensibly, with the wiring at the back.

These examples are hard to reconcile with the notion that each species was carefully designed by a deity. They remind me more of the plumbing system in an old house that has been repeatedly remodeled by a succession of plumbers, none of whom thought much about the house as a whole. In my old house, this has led to a plumbing system in which everything works, but nothing works as well as it might. There are toilets that stop up, faucets with too little pressure, and faucets where one must wait a long time for the hot water. When you take a wall apart, you find pipes that lead nowhere and pipes that reach their destinations by improbable routes. It looks a lot like the plumbing of the human body.

The paragraphs above describe several examples in which evolution arrived at a suboptimal design. Why does evolution not fix them? Because doing so by gradual steps would require evolving through intermediate forms that were even worse. It would require, in other words, evolving through adaptive valleys. For example, to unhook the vas deferens from the ureter, the testes would have to travel back into the interior of the body, across the ureter, and back out again. Along most of this route, the testes would be too warm to produce sperm efficiently. The individuals involved would have low fertility and would be selected against. This is just another way of saying that this evolutionary path goes across a valley in the fitness surface. We are stuck with design flaws because such adaptive valleys are hard to cross.

From this, we can draw two conclusions. First, it seems clear that organisms evolved. Had they been specially created by a deity, we would not expect design flaws. Yet flaws make perfect sense under the hypothesis of evolution. These design flaws also justify a second conclusion—that the fitness surface has lots of peaks and valleys. If the fitness surface were less rugged, there would not be so many places to get stuck.

Adaptive valleys may be hard to cross, yet it would be rash to conclude that evolution never crosses them. After all, some may be easier to cross than others. Let us look at a few examples.

Crossing adaptive valleys

One example involves snails of the South Pacific. If you hold a snail shell with the point toward your eye, the shell will spiral away from you. The shell is "right-handed" if these spirals are clockwise but "left-handed" if they are counter-clockwise. Right-handed and left-handed shells work equally well, but there is trouble when the two types try to mate. They often cannot bring their genital slits into contact. Neither snail is fertilized and neither reproduces. This problem is especially acute in species with wide, round shells [46]. For this reason, it is good to have a shell that coils in the same direction as those of the other snails around you. In a population of right-handers, a left-handed snail is a lonely snail: unable to mate and with low Darwinian fitness. Selection favors the common type, whichever type that is.

Because of this disadvantage, new mutations that change the direction of coiling are unlikely to spread. Every mutation is initially rare, and while rare it suffers a handicap. Thus, populations of right-handed snails tend to stay right-handed, and vice versa. In this sense, right-handers and left-handers constitute separate peaks on a fitness surface. Selection tends to prevent populations of snails from switching peaks.

Yet somehow, they occasionally manage to do so. In one case, we got a glimpse of the process in action. The island of Moorea lies just northwest of Tahiti in the south Pacific. Thirty years ago (before they were all driven extinct), it was home to several endemic land snails of the genus *Partula*. Most of these species were right-handed, but a few were left-handed. Since these species were all closely related, some species must have switched their coil fairly recently. But how? Bryan Clarke and James Murray found a clue in the distribution of one species, which had individuals of both coils [19]. *Partula suturalis* was left-handed in the west of its range but right-handed in the east. These regions were separated by a narrow transitional zone, about a kilometer wide, that contained both types. Clarke and Murray noticed an interesting coincidence: *suturalis* was right-handed only in that portion of its range that it shared with a closely related snail, *P. mooreana*. This second species happened to be left-handed. Where the two types coexisted, left-handed *suturalis* wasted time and resources in fruitless hybrid matings with *mooreana*. Because of these fruitless matings, left-handed *suturalis* individuals were at a real disadvantage. They had low fertility in regions dominated by *mooreana*. Clarke and Murray suggest that this may explain, at least in part, why this population switched from one adaptive peak to another—from left-handed to right-handed.

This seems plausible, but it is probably not the whole story. After all, rare left-handers would *still* have had trouble finding mates. The need to avoid hybrid matings might have made the valley shallower, but it would not have eliminated the valley altogether. Another factor may have made it shallower still. In these snails, the direction of coil is governed by an unusual genetic system. A snail's coil is determined not by its own genotype but by that of its mother. The very first right-handed mutant would have had a left-handed shell and would thus have had no trouble finding mates. Right-handed snails would begin to appear only when the daughters of this initial mutant had offspring of their own. These would still have been rare, but they would not have been unique. They could have mated with each other.

Both factors would have tended to make the adaptive valley shallower. Nonetheless, most experts think there was still an adaptive valley to cross. If so, these snails provide the best example from nature of an adaptive peak shift. However, we cannot be certain that the adaptive valley was really a valley.

Perhaps the strange genetics and the need to avoid hybrid matings gave right-handers an absolute advantage over left-handers. If so, then the snails could have shifted state without crossing any adaptive valley. No one is really sure. There are other examples of probable peak shifts in natural populations, but none of them is entirely convincing. All can be challenged in one way or another [22]. To find a really convincing case, we need to go to the laboratory.

As we saw in Chapter 2, it is hard to do laboratory experiments on the formation of new species. The trouble is that new species arise only rarely. You need to study a large population for many generations, and you are not likely to succeed in this if you study elephants. Thus, scientists have concentrated on species that have short generations and can be raised in large numbers. The same considerations also apply to the problem of crossing adaptive valleys. Christina Burch and Lin Chao decided to attack that problem using what is surely the smallest and fastest-breeding species that one can study [12]. They studied a virus that infects bacteria. Such viruses are called "bacteriophages," or more briefly "phages."

Their experiment had two phases. The first phase involved a series of "bottlenecks," or reductions in population size. In each bottleneck, the phage population was reduced to a single individual and then allowed to recover. (This didn't take long: in five generations there were several billion phage.) This process was repeated again and again: bottleneck, then recovery, then bottleneck again, and so on. Severe bottlenecks such as these cause large random changes in gene frequencies. These random shocks were so severe that they overwhelmed the much smaller changes made by selection. In this environment, selection cannot be effective. Within a few dozen repetitions, a harmful mutation had arisen and spread through the population. All the phage carried it, and it greatly reduced their fitness.

At that point, Burch and Chao shifted to the second phase of the experiment. The second phase was a lot like the first, except that the recoveries did not start from single phage. They started from large numbers, so that the population never passed through any narrow bottleneck in population size. Gene frequencies no longer bounced around at random, and selection was far more effective. It eliminated any harmful mutations that arose, and it encouraged the helpful ones. Gradually, the population regained its original fitness.

Although the loss of fitness happened in a single large step, the gain involved several smaller steps. This shows that the beneficial mutations did not simply reverse the mutation that caused the original damage. (Had they done so, fitness would have been restored just as it was lost—in a single step.) This implies that the phage population had a different genotype at the beginning of the experiment than at the end. Both of these genotypes had high fitness, although the route from one to the other involved intermediates with low fitness. In other words, Burch and Chao had watched their population evolve across an adaptive valley.

This experiment shows that no miracle is needed to push a population across an adaptive valley. Burch and Chao did it by repeatedly reducing the size of their population. There is nothing miraculous in that. Natural populations often suffer through harsh conditions that greatly reduce their numbers. These natural bottlenecks have effects analogous to those in the experiment, causing gene frequencies to bounce around at random. If these random shocks are large, they can overwhelm the effect of selection and push the population into an adaptive valley.

Nonetheless, there are reasons to be skeptical that peak shifts are important in nature. The bottlenecks of Burch and Chao were *very* severe, reducing the population size to a single individual. When this happens in nature, the population seldom survives. For more realistic population sizes, it can take a very long time to cross an adaptive valley [22]. On the other hand, the whole process speeds up if the fitness surface wobbles a little [119], or if some populations are more isolated than others [83], or if the population is continuously distributed in space [5], or if the population's size varies [118]. Do these factors speed things up enough to make peak shifts a plausible engine of adaptive evolution? Maybe, but no one knows for sure.

Where does this leave us? The many examples of poor engineering in nature imply that species *do* get stuck on suboptimal adaptive peaks. This shows that the adaptive landscape is rugged, with lots of peaks and valleys. It is also clear that evolution *can* cross adaptive valleys, at least in the laboratory. It is not clear whether this ever happens in nature. There are a few plausible cases, but only a few, and none of these is beyond challenge.

For these reasons, evolutionists disagree about the importance of peak shifts in adaptive evolution. Some argue that peak shifts play no central role in adaptive evolution. As Richard Dawkins puts it, "there can be no going downhill—species can't get worse as a prelude to getting better" [32, p. 91]. Others are not so sure. This is an open question in evolutionary biology.

Conclusion

This chapter began with Denton's argument that natural selection cannot build complex adaptations because doing so would require crossing adaptive valleys. This (according to Denton) would require a miracle; it is something that evolution cannot do. We are at last in a position to evaluate that argument.

First, let us agree that Denton was right about one thing: the adaptive landscape is rugged, with lots of peaks and valleys. There is no other way to interpret the extensive evidence of poor design in nature. It does not follow however that adaptive evolution requires crossing valleys. We saw in Chapter 4 that complex adaptations can evolve via a series of small, individually advantageous changes. No valley need be involved. Many evolutionists would argue that this is the whole story—that all adaptations evolve in this way. If this view is correct, then the adaptations that exist are simply the ones that could be reached *without* crossing valleys.

On the other hand, we need not appeal to miracles even if evolution does cross valleys. Denton was wrong about this too. In small populations, gene frequencies are buffeted by a variety of random forces, and these can push populations across valleys. We understand the mechanisms involved, and we have seen them operate in the laboratory. Evolutionists may argue about how often they happen in nature, but one thing is clear: there is no plausible basis for the argument that adaptive evolution requires miracles.

6

ISLANDS IN THE 21ST CENTURY

T his book began with Darwin on the Galapagos Islands, puzzling over the distribution of plants and animals there. It seemed strange that each island had its own species of mockingbird. Why would the creator put different species on such similar islands? And if he were so inclined, why make them all mockingbirds, similar not only to each other, but also to those Darwin had seen on the South American mainland? Why was this pattern repeated in the finches, the tortoises, the insects, the lizards, and even the plants? And finally, why were the islands inhabited exclusively by species capable of long ocean journeys? These were the facts that first prodded Darwin to think the unthinkable—that the diverse Galapagos species had not been created independently but had evolved. This theory, as Darwin realized, solved all of these puzzles at a single stroke.

Yet in a sense, Darwin had it far too easy. His theory implied other predictions that Darwin had no way of testing. For example, there ought to be a relationship between the phylogenetic tree of the birds and the history of the islands.[1] At a minimum, the original pair of mockingbirds must have arrived long ago, and they must have arrived on an island that was already in existence. Thus, the deepest node in the birds' phylogeny should separate birds on one of the older islands. As each new island appeared above the waves, it would have been colonized by immigrants from other islands. This implies a close connection between the phylogeny and the history of island formation. I'll explain that connection in a moment, but first you need to know a few of the things that Darwin didn't know.

---

[1] Phylogenetic trees are introduced on pp. 25–31.

What Darwin didn't know about volcanic islands

Ever since the 19th century, geologists have wondered why the surface of the earth is lumpy rather than smooth—why there are continents and ocean basins. Geologists quarreled about this during the 19th century, and they were still quarreling during the 1950s. In the 1960 edition of the *Encyclopædia Britannica*, Philip Lake summarized the theories still in play. One held that the earth's skin wrinkled as it cooled. Another held that the continents occupied fixed positions, but portions of the ocean basins were from time to time elevated to form "land bridges" that stretched from continent to continent. Lake was dismissive with regard to the radical idea of continental drift, which held that continents consist of lighter material that "floats" on a sea of heavier rock—that they occupy no fixed positions but drift about on the surface of the earth. Lake described one aspect of this theory as "fanciful." In the past quarter century, he felt, the idea of the permanence of the ocean basins had begun to gain ground [66, p. 335].

He could not have been more wrong. Ten years later, new geophysical discoveries had established the reality of continental drift beyond all dispute. When I took geology in 1970, continental drift had just made it into undergraduate textbooks, and I was taught nothing about the ideas that had seemed so important to Philip Lake just 10 years before.

By 1970, geologists were busy reinventing their discipline. One aspect of this revolution involved chains of volcanic islands. One has only to look at a map of the Pacific Ocean to realize that something strange is going on. The islands are not scattered at random but are grouped into chains, each strung out in a relatively straight line. In the south Pacific, these are nearly parallel, with each chain running from northwest to southeast. Take for example the Hawaiian Islands, which are shown in Figure 6.1. The largest island (itself called "Hawaii") is at the southeast end of the line. To the northwest, the islands get progressively smaller. Only the major islands are shown in the figure, but the chain continues hundreds of miles farther as a series of small islands and coral atolls. Even beyond the last island, it continues as a series of ridges and seamounts under the sea. Altogether, the chain reaches nearly to Russia's Kamchatka Peninsula, 6,000 kilometers to the northwest.

The only island with active volcanoes is the largest one—Hawaii—at the southeast end of the chain. Radiometric dating shows that it emerged from the sea roughly 0.4 million years ago. To the northwest along the chain, each island is older than the last. At the northwest end lies the oldest major island, Kauai, which is just over 5 million years old. Far to the northwest and deep beneath the sea, the oldest part of the chain is 82 million years old.

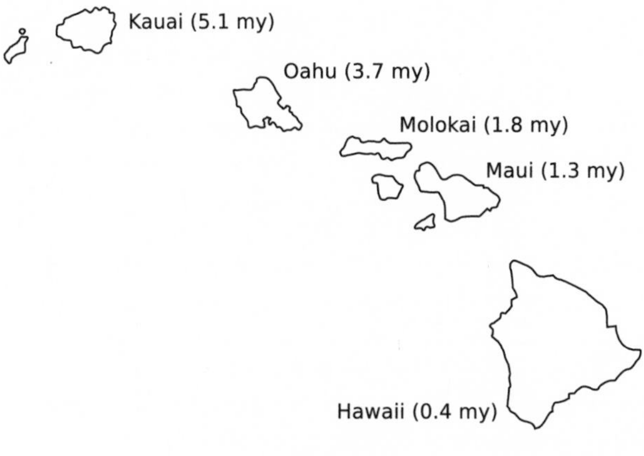

Figure 6.1: Hawaiian island chain, with names and ages of major islands [95].

To understand these islands, you need to know that it is not only continents that drift. The earth's crust is broken into gigantic plates, which float on its surface. The Hawaiian Islands lie on the Pacific Plate, which drifts slowly to the northwest at roughly 10 centimeters (4 inches) per year. Deep inside the earth, a plume of heat rises from the earth's molten core to produce a "hotspot" near the surface. This hotspot gives rise to volcanoes. Each island in the Hawaiian chain had its turn over the hotspot, and at that time its volcanoes were active. Eventually however the volcanoes on each island go extinct as the island drifts beyond reach of the hotspot. The hotspot is currently under the island of Hawaii, the only island in the chain with active volcanoes.

All of the island chains in the Pacific Ocean have similar histories, including the Galapagos. Yet none of this was understood until the 1960s. Over a century earlier, Darwin knew none of this and was therefore in no position to test hypotheses about the connection between phylogeny and geological history. But that was not his only problem. He was nearly as ignorant about phylogeny.

What Darwin didn't know about phylogeny

Without an accurate phylogeny, the evolutionist confronts a world of incomprehensible mysteries. Take for example the anteaters. On both sides of the South Atlantic, there are toothless creatures with long snouts and even longer tongues, which they use to probe deep into ant hills and termite burrows. The early comparative anatomists grouped them into a single order, *Edentata*. This posed a problem: How did the edentates get across the Atlantic Ocean? They

65

certainly could not have swum. Was there at one time an isthmus that stretched all the way from Africa to South America? Some thought so, but this led to other problems. There are also similarities connecting other pairs of continents, and each of these seemed to require its own isthmus. As the number of isthmuses multiplies, the hypothesis begins to look absurd.

This problem was caused by a mistaken idea about phylogeny. As we now know, there is no close relationship between Old World and New World anteaters. Their similarities evolved independently, as each group specialized on foraging for ants and termites. The story illustrates the central role of phylogeny in any effort to understand how plants and animals are distributed across the landscape. Unfortunately, it is easy to get phylogeny wrong.

During all of the 19th and much of the 20th centuries, inferences about phylogeny were based mainly on morphological comparisons and did not involve any systematic method. Instead, the problem was seen as subjective, and practitioners relied heavily on intuition. It is no wonder that the phylogenies of the day were unreliable. The situation hadn't changed much in the early 1960s as biologists began trying to exploit the new insights of continental drift. Things were not altogether hopeless, for biological data are so obviously hierarchical that any sensible method will produce a tree that is correct in broad outline. This was enough, in the 1960s and 1970s, for biologists to solve many outstanding problems in biogeography. For the most part, those solutions still stand. Yet it has turned out that the trees of those decades were incorrect in many details, and this made it hard to relate phylogeny to evolutionary history.

All this was about to change. During the 1960s, biologists began describing organisms in terms of their DNA and proteins, and the era of molecular evolutionary genetics was born. These new data spurred others to develop new and more systematic methods for inferring phylogenies. Since the 1960s, this process has continued, aided by steady increase in the power of computers. The trickle of papers on molecular phylogeny that began in the 1960s has become a flood, and we now understand phylogenetic relationships far better than ever before.

It is not that molecules have made morphology irrelevant, for the two are complementary. Molecular data are useful in figuring out relationships among modern species, and they even help to figure out when their common ancestors lived. They tell us little however about what that ancestor looked like or how it lived. For that, we rely on morphology and (with luck) fossils. All three categories of data have been useful in unraveling the thorny knots of biogeography.

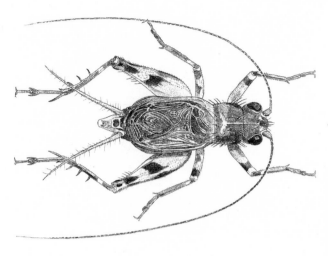

Figure 6.2: The Hawaiian flightless cricket, *Laupala nigra*. From Fig. 84, p. 131 of reference [85] by permission of Daniel Otte.

Hawaiian flightless crickets

It is time to return to our real interest, the relationship between phylogeny and the geological history of islands. I started this chapter with mockingbirds, but they are not the ideal species for our purposes. The trouble is that they fly, and this makes it easy for them to hop from island to island. The more of this they do, the more the populations on different islands mix, and the less will their phylogeny reflect the history of the islands. We will be better off with a species that only rarely disperses from island to island. The Hawaiian Islands are home to hundreds of such species.

If you venture into a tropical rain forest there, be quiet and listen. You may hear a delicate tinkling sound that seems to come from the leaf litter, the tangles of ferns, and the forest undergrowth. The source of this sound is hard to find. It is not the snails though you will see them everywhere. It is crickets of the genus *Laupala* [85, p. 129]. As you can see from Figure 6.2, these crickets lack wings and cannot fly. Perhaps for this reason, they rarely move from one island to another. Let us consider how their phylogeny might be expected to look, in view of the history of the islands on which they live.

The Hawaiian Islands were formed one at a time, and each island is a little younger than the one just to the northwest. Before discussing real crickets, let us consider how this sort of history might affect phylogeny in an ideal world. To keep things simple, I will boil the cricket phylogeny down to a condensed version called an *area cladogram*. In such a graph, there is a branch tip for

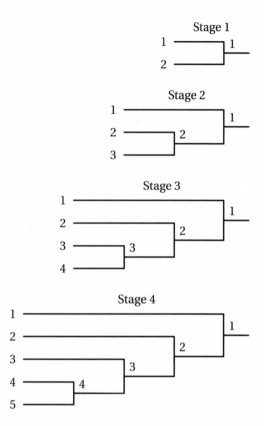

Figure 6.3: How a sequence of island colonizations generates a stair-stepped area cladogram. Numbers represent islands, from oldest to youngest. At each stage, a new island is colonized by immigrants from the next-youngest island, and the branch leading to that next-youngest island splits in two.

each island rather than for each species. In addition, I number the islands from oldest to youngest and use these numbers (rather than island names) as labels.

Figure 6.3 shows how a hypothetical area cladogram might develop over time. In Stage 1, only the first two islands exist, and the cladogram has just two branches. When island 3 arises, it is colonized by immigrants from the nearest island. The nearest island is always the next-youngest, which in this case is island 2. Thus the new species that evolve on island 3 are most closely related to species on island 2. This fact is reflected in the phylogeny, where branch 2 splits to form (2, 3). As each new island arises, this process repeats: branch 3 splits to form (3, 4), and so on. In the ideal case, this process produces an area cladogram that is stair-stepped, as shown in Figure 6.3.

In the real world, things are seldom so perfect. Some islands are colonized from the next-nearest island rather than the nearest, and some islands are

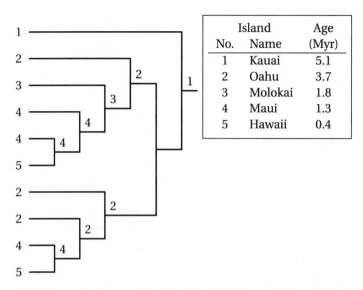

	Island	Age
No.	Name	(Myr)
1	Kauai	5.1
2	Oahu	3.7
3	Molokai	1.8
4	Maui	1.3
5	Hawaii	0.4

Figure 6.4: Area cladogram for Hawaiian flightless crickets [69]. Numbers refer to islands, as in key. Branch tips may refer to several related species.

colonized more than once. Nonetheless, there is often at least a trace of the stair-stepped pattern seen in Figure 6.3.

So much for theory; now for real crickets. Kerry Shaw and Tamra Mendelson have been studying *Laupala* crickets for years. They collected genetic data from crickets throughout the islands and estimated their phylogenetic tree using a method similar in spirit to the one we used in Chapter 3. After boiling their phylogeny down to an area cladogram, we arrive at Figure 6.4. The deepest node there separates Kauai from the other islands. So far so good—this is exactly like the deepest node in the ideal case. After that, the cladogram gets a little more complicated. The non-Kauai portion falls into two sub-trees. This suggests that the Oahu crickets split into at least two species, *both* of which colonized the younger islands independently. This does not contradict the historical story, but it does require that we examine the two sub-trees separately. (After all, they represent independent evolutionary experiments.) The upper sub-tree implies that islands were colonized in the sequence $2 \rightarrow 3 \rightarrow 4 \rightarrow 5$, precisely as in the ideal case. In the lower sub-tree, the sequence is $2 \rightarrow 4 \rightarrow 5$. Within each sub-tree, the sequence of implied dispersal events is always from older islands to younger ones, in perfect agreement with the stair-stepped pattern.

What can we conclude from all this? First realize that every one of these islands has several *Laupala* species, none of which live on more than one island. The area cladogram implies that all these species evolved from a single ancestor, which lived on Kauai (or possibly some earlier island now beneath the sea). This conclusion is supported first by the genetic data. This support is strong because of the large amount of genetic data used. Nonetheless, one might quibble that this doesn't count because the cladogram was carefully *fit* to the genetic data. However, the cladogram also agrees perfectly with the history of island formation. This is so moreover in two independent sub-trees. If the cladogram were a statistical artifact, we would not expect it to be consistent in detail with totally independent evidence. Yet it is, and this fact is powerful support for the validity of the cladogram. This in turn provides powerful support for the view that evolution has happened on these islands. In fact, the rate at which these crickets have formed new species is the highest ever recorded in an arthropod.

Conclusion

We are all good at making up explanations to fit the facts that we see. Yet even if it does fit the facts, our explanation may still be incorrect. For this reason, scientists like to test theories against facts that had *not* been seen by the author of the theory. The *Laupala* crickets provide such a test. They test Darwin's theory against two classes of fact—geological history and molecular phylogeny—that were unknown in the 19th century. As we have seen, Darwin's theory predicts extremely well, even with experimental results that Darwin could never have imagined.

7 HAS THERE BEEN ENOUGH TIME?

It may be objected, that time will not have sufficed for so great an amount of organic change, all changes having been effected very slowly through natural selection.

Charles Darwin, 1859 [27, p. 282]

So far as this world is concerned, past ages are far from countless; the ages to come are numbered; no one age has resembled its predecessor, nor will any future time repeat the past. The estimates of geologists must yield before the more accurate methods of computation, and these show that our world cannot have been habitable for more than an infinitely insufficient period for the execution of Darwinian transmutation.

Fleeming Jenkin, 1867 [59, pp. 320–321]

E volution works by small changes accumulated over many generations. And if evolutionists are right, this plodding process generated every living thing—from viruses to whales, from humans, to trees—all starting from a single ancestor. This must have taken time, an immensity of it. But has there really been so much time? Is the earth really that old?

From the quotes above, it is clear that Fleeming Jenkin didn't think so and that Darwin was worried. In the same passage, Darwin went on to say that "he who...does not admit how incomprehensibly vast have been the past periods of time, may at once close this volume."

In the 19th century, there were two reasons for skepticism about the idea of an old earth. First, biblical scholars of the 17th century had arrived at very modest estimates of the earth's age. The most famous of these was published in 1658 by James Ussher, Anglican Archbishop of Armagh, Ireland. According to his chronology, the beginning of the world

71

fell upon the entrance of the night preceding the twenty third day of Octob[er] in the year of the Julian [Period] 710. The year before Christ 4004. [109]

The earth was thus about 6,000 years old. Ussher's estimate has been enormously influential over the years and underlies the "young earth" creationist movement of today.

But Jenkin was skeptical for a different reason. His background was in engineering, and he had since 1857 been working on the great engineering project of the day: laying telegraph cable across the Atlantic Ocean. Also working on that project was the physicist William Thomson (later Lord Kelvin). Thomson and Jenkin formed a friendship that was to last their whole lives. Jenkin was thus favorably disposed to his friend's ideas about the age of the earth.

Thomson was interested in heat and had collated data about temperature within deep mines. The deeper you went, the warmer it got. For Thomson, this could mean only one thing: the earth had been warmer in the past and was gradually cooling. Thomson worked out an equation describing this process and with it was able to calculate the earth's temperature at any given time in the past. If you went back far enough—about 100 million years—the earth was so hot that the rocks would melt [108]. This is the work that Jenkin was referring to in the quote above.

It is easy to show that something is wrong with Ussher's calculations. There are several kinds of annual layer to be found in nature, and one can measure time simply by counting layers. For example, one can count tree rings back to 11,855 years before present (BP) [64]. Another kind of annual layer accumulates beneath some lakes and shallow ocean basins. The sediment there forms annual layers called *varves*. These have been counted back to 37,930 BP [61]. The Green River formation of Wyoming and Utah has several million varves, which were deposited in an ancient lake [11]. The earth is clearly much older than 6,000 years. Thomson's calculations had a problem too. They were based on temperatures within deep mines, and those temperatures were elevated by a phenomenon that was then unknown: radioactivity. Because of this effect, Thomson's data are uninformative; the earth may be a lot older than he thought. But how much older?

The answer to this question also depends on radioactivity, but in an entirely different way. Not only do radioactive minerals release heat, they also decay at a constant clock-like rate. If you know how much of the mineral was there to start with and how much is there now, it is easy to figure out how much time has elapsed. The first section below explains how this works. The second deals with an obvious problem: we seldom know how much of the mineral was there when the rock was formed.

Radiometric clocks

The chemical properties of an atom determine the ways in which it can bond with other atoms. For each atom of oxygen (or carbon, or helium, or any other element), these chemical properties are the same. That is why they are said to belong to the same element. But atoms of a given element may differ in weight, and these different forms are called *isotopes*. Physicists prefer to speak of *mass* rather than *weight*. (The mass of an object measures how hard it is to stop once in motion.) Each isotope is thus referred to by its *atomic mass*.

For example, most atoms of oxygen have atomic mass 16 and are called "oxygen 16" (^{16}O). A few have atomic mass 18 and are called "oxygen 18" (^{18}O). These isotopes are *stable*, which means that they have no tendency to change into anything else. We are interested here in other isotopes that are not stable: they are *radioactive*.

Radioactive isotopes are unstable. An atom of such an isotope (called the parent) will eventually decay into something else (the daughter). No one can predict when this will happen, but the risk of decay is constant in time. For example, the most common isotope of uranium (U) is ^{238}U. This isotope is radioactive and will eventually decay into a daughter atom of ^{206}Pb, a stable isotope of lead (Pb). This is a very slow process: in a large sample of ^{238}U atoms, half will decay in 4.46 billion years. This interval is called the *half-life* of ^{238}U. Other radioactive isotopes decay more rapidly and have shorter half-lives. Still others decay even more slowly than ^{238}U.

Figure 7.1 tracks a sample of ^{238}U atoms through time. Their number very gradually decreases, and for each one lost we gain an atom of ^{206}Pb. The total number of atoms thus stays the same. Other radioactive isotopes decay according to similar curves but with different half-lives.

How might this curve be used to date rocks? We can measure the amount of the parent (^{238}U) and daughter (^{206}Pb) in a rock today, but this does not tell us what fraction of the parent has been lost. To discover that, we would need to travel back in time and examine the rock at the time it formed. Suppose for the moment that we accomplished this and learned that a fraction 0.627 of the original parent was still there. Find 0.627 on the vertical axis of Figure 7.1 and follow the dotted line. You will find yourself on the horizontal axis at 3 billion years. This is the age of the rock. On the other hand, if half of the original parent were gone, we would conclude that the rock was 4.46 billion years old—one

73

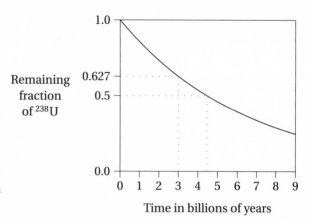

Remaining fraction of ^{238}U

Figure 7.1: Decay of uranium (^{238}U) atoms into lead (^{206}Pb).

Time in billions of years

half-life. This principle underlies all radiometric dating. It is the principle of the radiometric clock.

Objections spring immediately to mind. Can we really know how much parent and daughter were in the original rock? What if some were lost or added? For that matter, how do we know the half-life is constant or that it equals 4.46 billion years? Let us take these objections in turn.

The last objection seems fatal. How can one measure a rate so slow that its half-life is measured in billions of years? The answer however is easy. When a radioactive atom decays, it emits a particle that can be detected by an instrument such as a Geiger counter. The rate of decay is estimated from the rate at which particles are emitted by a known quantity of uranium. You might think that with a half-life of 4.46 billion years, such experiments would take a long time. However, tiny quantities of uranium contain such huge numbers of atoms that there is no such difficulty. In a single gram of uranium, about 10,000 atoms decay per minute [62].

How do we know the rate is constant? First, we can watch the entire process using isotopes that decay rapidly. This is what originally led physicists to the view that rates are constant [97]. Second, this constancy makes sense. The forces that hold a nucleus together are a million times stronger than those involved in chemical interactions. To change the rate of radioactive decay, one needs energies such as those generated within nuclear bombs and stars. Rocks on the earth and other planets do not experience such energies [26, p. 87]. Thus, the constancy of radioactive decay makes theoretical sense. It is wise however to be skeptical of theories. For this reason, physicists have exposed radioactive isotopes to all manner of extreme conditions, none of which changed any rate of decay by as much as a percent [26, p. 89].

There is a lot of evidence here, but it is a bit circumstantial. It would be far more satisfying if we could somehow show that rates of decay have been constant over thousands or even millions of years. Although this sounds impossible, it has already been done.

Before explaining this remarkable feat, I must deal with the other objections listed above: how can we tell how much of the parent and daughter isotopes were in the original rock or whether any has been gained or lost? These problems are dealt with in various ways by different methods of radiometric dating. We will consider just one—the method of isochrons—which is used with various isotopes including those we have been discussing.

Isochrons

We will study three samples from a hypothetical igneous rock. Since this is only an exercise, the rock's age will be known in advance. As we shall see, this known age can be recovered by constructing a graph. For simplicity let us assume the rock's age to be 4.46 billion years, the half-life of ^{238}U.

An igneous rock contains grains—small crystals—of a variety of minerals. Our samples consist of three such grains, each chosen from a different type of mineral. Within these samples, we will be interested in three isotopes. Two of them (^{238}U and ^{206}Pb) were discussed above. The third, ^{204}Pb, is useful because its quantity does not change. None is lost because ^{204}Pb is a stable isotope with no tendency to decay into something else. None is gained because it is not produced by any form of radioactive decay. This makes it useful as a sort of baseline.

In our imaginations, let us travel back to the time 4.46 billion years ago, when the rock was formed. At that earlier time, we count the number of atoms of each isotope in each of the three mineral grains. These counts (which in reality I just made up) are shown in the upper panel of Table 7.1. To obtain the lower panel of the table, we travel back to the present and count the atoms once again. Note that in getting from the upper panel to the lower, half the atoms of ^{238}U decay into ^{206}Pb, reflecting our assumption that the age of the rock equals the half-life of ^{238}U. Note also that the number of ^{204}Pb atoms does not change.

Columns E and F show the ratios ^{238}U/^{204}Pb and ^{206}Pb/^{204}Pb. These values are plotted against each other in Figure 7.2. The points in the graph represent the three samples in the initial rock (○) and in the modern rock (●). All the points fall onto two straight lines, one for the ancient rock and one for the modern rock. Each straight line is called an *isochron*. The isochron is horizontal for the initial rock but slopes at 45° for the modern rock. In general, the slope of an isochron reflects the age of the rock. This is why isochrons tell us about age.

(a) 4.46 billion years ago

A	B	C	D	E	F
Samp.	^{238}U	^{206}Pb	^{204}Pb	$\dfrac{^{238}\text{U}}{^{204}\text{Pb}}$	$\dfrac{^{206}\text{Pb}}{^{204}\text{Pb}}$
1	80,000	20,000	1,333	60	15
2	30,000	10,000	667	45	15
3	90,000	50,000	3,333	27	15

(b) Today

A	B	C	D	E	F
Samp.	^{238}U	^{206}Pb	^{204}Pb	$\dfrac{^{238}\text{U}}{^{204}\text{Pb}}$	$\dfrac{^{206}\text{Pb}}{^{204}\text{Pb}}$
1	40,000	60,000	1,333	30.0	45.0
2	15,000	25,000	667	22.5	37.5
3	45,000	95,000	3,333	13.5	28.5

Table 7.1: Samples from 3 mineral grains in a hypothetical rock. The two tables describe the rock as it existed 4.46 billion years ago (a) and today (b). Columns B–D show the number of atoms of each isotope; columns E–F show isotope ratios.

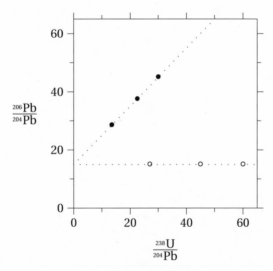

Figure 7.2: The plotted values represent samples as they existed in the initial rock (○) and in the modern rock (●). The straight lines are called *isochrons*.

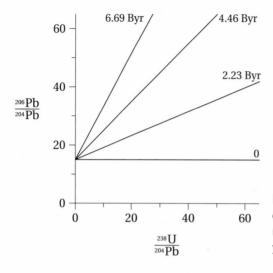

Figure 7.3: Isochrons of different ages, with age measured in billions of years (Byr).

The older the rock, the greater the slope of the isochron. This happens because with each passing year, we lose some atoms of ^{238}U (moving us left on the graph) and gain some of ^{206}Pb (moving us up). As each point on the isochron moves up and to the left, the slope increases. This process is illustrated in Figure 7.3. When the age of the rock equals the half-life of ^{238}U, the slope is exactly 45° as shown in Figure 7.2. Isochrons let us date rocks even when the isotopic composition of the original rock is unknown.

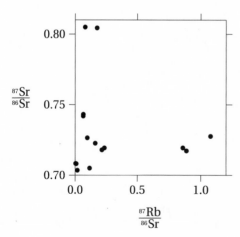

Figure 7.4: Rubidium-strontium data from a metamorphic rock in the Panamint Mountains of California [114]. The samples do not lie on an isochron, so the rock can't be dated.

77

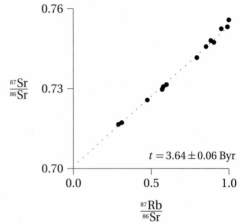

Figure 7.5: Rubidium-strontium isochron for Greenland gneiss [26, p. 149]. Note that axes differ in scale.

Occasionally the method fails. This is illustrated in Figure 7.4, which plots ratios of rubidium and strontium isotopes in a real metamorphic rock from California. The isochron method fails here, apparently because this rock was exposed to intense heat that allowed atoms of some isotopes to escape. But since heating affects some minerals more than others, the loss was not evenly distributed among samples. They no longer lie along an isochron, and the rock cannot be dated. It is reassuring that even when its assumptions fail, the method does not mislead us. It simply gives no answer at all.

Fortunately, isotope ratios often do lie on isochrons. Figure 7.5 plots samples from a piece of metamorphic rock (gneiss) from Greenland. The data points are all very close to the straight line, indicating that the assumptions hold. The slope implies an age of 3.64 billion years. In such cases, the method's internal checks and balances give reason for confidence in the results.

There are also other grounds for confidence. Radiometric methods give the right dates for historical events, such as the eruption of Mount Vesuvius in 59 BC [91]. They have been calibrated with success against tree rings [64] and varves [61] going back 45,000 years. In older strata, different methods can be checked against each other and typically give very similar dates [92].

Perhaps the most impressive test involves Paleozoic corals. To understand this analysis, let us start with the tides. Tides roll in and out twice a day all over the earth under the combined influence of the earth's rotation and the moon's gravity. All this motion causes friction, which is gradually slowing the rotation of the earth. Millions of years ago, the earth turned faster, and there were more days in a year. Astronomers tell us that 345 million years ago, in the

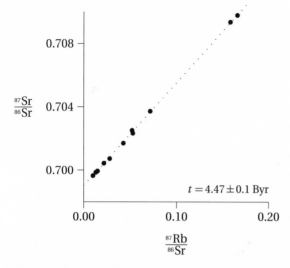

$t = 4.47 \pm 0.1$ Byr

Figure 7.6: Rubidium-strontium isochron for lunar rock [26, p. 248]. Note that the axes differ in scale.

Devonian period, days were 22 hours long, and there were 396 days per year. We can test this prediction because some Devonian corals put on daily rings that were thicker in summer than in winter. Sure enough, this "coral clock" shows about 400 days per year [116]. This agreement is remarkable. It is based on calculations from three disciplines—astronomy, geology, and biology—that involved completely different theories and types of data. Had any of these been wrong, the two numbers would not have agreed. Yet they do, and this fact shows that the radiometric dates are accurate far back into the past.

The age of the earth

There are three sources of information about the age of the earth: rock from the earth, from the moon, and from meteorites. We have already seen a 3.64 billion year date from the earth (Figure 7.5), but it is far from the oldest. The oldest thus far is 4.4 billion years [122]. The earth however is not an ideal place to search for very old rock. Its molten core generates movement of its crust, which we observe in earthquakes and volcanoes. As these forces push new rock up onto the surface, they also suck old rock into the earth's interior. Because of this recycling, little or none of the earth's original crust remains.

This recycling stopped long ago on the moon, whose core is no longer molten. The moon is thus a better place to search for very old rock. This rock is relevant to us because it is likely that the earth and moon formed at the same time. Many samples of moon rock were brought back to earth by the

79

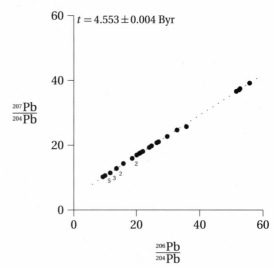

Figure 7.7: Lead-lead isochron for the Allende meteorite [26, p. 327]. Numbers show the number of data points covered by the adjacent symbol.

Apollo program of the late 1960s and early 1970s. These date to a variety of ages, ranging back to 4.47 billion years, as shown in Figure 7.6.

Figure 7.7 shows an even earlier isochron obtained from a meteorite. These lead isotopes are especially important because the earth's lead lies on the same isochron with many meteorites [76]. This provides strong evidence that the earth and the meteorites formed at about the same time. The 4.55 billion year date shown in Figure 7.7 has been obtained from many meteorites and is the best estimate we have of the age of the earth.

Conclusion

The earth is many times as old as imagined by either William Thomson or James Ussher. In retrospect, it is easy to see where Thomson went wrong. Nothing was known in his day of radioactivity. His calculation thus failed to account for the radioactive heat generated within the earth. It is ironic that radioactivity played such contrasting roles in the 19th and 20th centuries. In the 19th century it was the unknown factor that spoiled efforts to estimate the age of the earth. In the 20th it was the key that enabled us to obtain that estimate.

We saw on p. 42 that a complex eye can evolve from an eye spot in about 360,000 years. The earth has existed long enough for 10,000 such eyes to evolve, one after another. There is simply no room to argue that evolution has had too little time.

8   DID HUMANS EVOLVE?

We must acknowledge, as it seems to me, that man with all his no-
ble qualities, with sympathy which feels for the most debased, with
benevolence which extends not only to other men but to the hum-
blest living creature, with his god-like intellect which has penetrated
into the movements and constitution of the solar system—with all
these exalted powers—Man still bears in his bodily frame the indeli-
ble stamp of his lowly origin.

Charles Darwin, 1871 [28, p. 405]

Even if other species did evolve, perhaps we are different. Perhaps we sprang directly from the hand of God. This point of view has a long history and made a certain amount of sense in the 19th century, when the evidence for human evolution was thin. But over the past century, that evidence has grown. In the past few decades, it has become overwhelming. Let us begin with fossils.

Fossils

When Darwin published his famous book, few authentic human fossils were known, and the science of paleoanthropology did not yet exist. Neither Darwin nor his contemporaries based the case for human evolution on fossils. They showed that humans resemble apes in minute detail, bone for bone, nerve for nerve, and blood vessel for blood vessel. What accounts for these commonalities, they asked, if not descent from a common ancestor?

In the decades that followed, fossils of archaic humans gradually accumulated, but progress in understanding them was slow. They were difficult to date, and this led to all sorts of confusion. For example, the Galley Hill fossil was a modern human skeleton that had been buried in ancient sediment. The excavators mistakenly concluded that the skeleton was as old as the surrounding sediment. The find seemed to show that humans of modern type had existed in Europe since early in the Pleistocene. There was also the Piltdown fossil, which later turned out to be a hoax. It consisted of a modern human skull and an orangutan jaw, which had been stained so their colors matched, filed so they fit together, and buried in an ancient gravel deposit. This hoax derailed the discipline for a generation. The confusion lasted until about 1950, when it was rapidly swept away by new dating techniques such as those described in Chapter 7. Since then, progress in paleoanthropology has been steady, and the field now provides some of the best evidence that humans evolved.

Before plunging in, let us discuss the sort of evidence that fossils might potentially provide. The section that follows this one will argue that humans are more closely related to chimpanzees than to any other species. This does not of course mean that we evolved *from* chimpanzees. It means that both species evolved from some common ancestor. Genetic evidence suggests that this ancestor lived about 6 million years ago. Between then and now, we might hope to find fossils that are in some sense intermediate between that ancestor and ourselves. What characteristics should such intermediates have?

Although the chimp-human common ancestor is not around for us to examine, we can make reasonable guesses. No living ape routinely walks on its hind legs. It is thus a good bet that this condition evolved in our own lineage at some time after the split with chimpanzees. We also differ from living apes in many other ways. Our brains are much larger, and we have vertical foreheads without heavy brow ridges. We have no bony ridge of muscle attachments running lengthwise along the tops of our skulls. Our canine teeth are barely longer than the others. This list is incomplete; it merely illustrates some of what we are looking for in intermediate fossils.

The relevant fossils are all *hominins*, a term that refers to humans and their ancestors up to but not including the common ancestor with chimpanzees. But it does not refer *just* to our ancestors. After the split with chimpanzees, the hominin lineage may at various times have split to form two or more *sibling species*, and these would be hominins too. If two hominin species coexisted, there is no guarantee that we will know about them both. Our fossil collections may include only one. Even if both are represented, we may be unable to tell their bones apart. For these reasons, we can never be certain that any particular fossil belonged to a species that was ancestral to modern humans. It *may* have done so, but on the other hand it may be a collateral relative—a member of some sibling species that later went extinct.

This fact is often cited by creationists who wish to dismiss the relevance of hominin fossils. Yet it was also true of the fossil whales discussed in Chapter 3, and in their case it did not seem problematic. To see why, let us return to *Rodhocetus*, the fossil whale on p. 21. It exhibited a suite of characters that in the modern world are found only in whales. These do not imply that *Rodhocetus* was ancestral to modern whales, but rather that it was related to them—that its ancestors were also theirs. On the other hand, *Rodhocetus* had legs, so its ancestors must have been land mammals. If both of these facts are true, then those land mammals must also have been ancestors of modern whales. Let me repeat: the ancestors of modern whales must have been land mammals. And we reach this conclusion without assuming anything about the descendants of *Rodhocetus*. In the same fashion, we can study fossil hominins without assuming that these particular fossils were ancestors of modern humans.

Let us begin with skulls. Figure 8.1 shows 16 of these, arranged in order of age. Take a moment and look for yourself. What changes do you see as you move backwards in time? The youngest skull (a) has the high forehead, globular braincase, and small brow ridges of a modern human. The oldest pair (o and p) show the sloping forehead, small braincase, and pronounced brow ridges of an ape. Along the way, we see many varieties of intermediate form. Skulls b–e are fairly modern yet have lower braincases and more pronounced brow ridges. By the time we reach skulls f–i, the brow ridges are large and ape-like, and the forehead is low, yet the braincase is still twice as large as that of any ape. Clearly, there are intermediate forms throughout the hominin fossil record.

Figure 8.2 makes this point more systematically by plotting the cranial capacities of hominin skulls against their ages. The earliest skulls, which are around 3 million years old, have cranial capacities close to 500 cc. This is similar to the size of a modern chimpanzee. In this respect, these ancient hominins were typical apes. Cranial capacity increases gradually with time. There are many intermediates between the size typical of apes and that typical of modern humans.

One might argue that before 2 million years ago, the creatures in the graph simply *were* apes. However, they walked on their hind legs just as we do. This point is more important than it might first appear. After all, ostriches and meerkats walk that way too, yet their anatomy is very different from ours. This is *not* the case for the early hominins. Their locomotor anatomy is human in countless small details—the structure of their feet, their knees, their pelvis, the way their spine connects with their skull. These detailed resemblances make sense only if we share a common ancestor with these creatures or if they were themselves our ancestors. And yet they have the brains of apes.

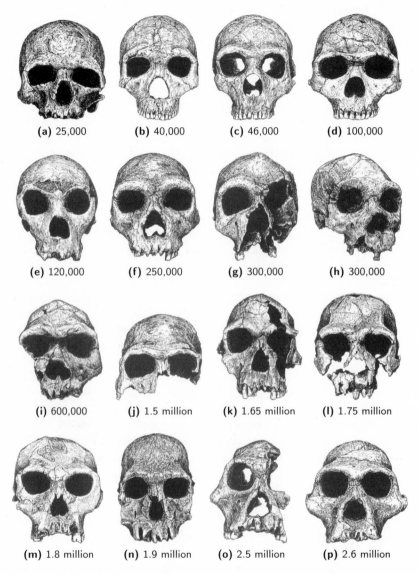

(a) 25,000   **(b)** 40,000   **(c)** 46,000   **(d)** 100,000

(e) 120,000   **(f)** 250,000   **(g)** 300,000   **(h)** 300,000

(i) 600,000   **(j)** 1.5 million   **(k)** 1.65 million   **(l)** 1.75 million

(m) 1.8 million   **(n)** 1.9 million   **(o)** 2.5 million   **(p)** 2.6 million

Figure 8.1: Fossil skulls, with ages in years. Reprinted by permission of Waveland Press, Inc. from C.S. Larsen, et al., *Human Origins: The Fossil Record*, 3rd edn. Waveland Press, Inc., Long Grove, Illinois, 1998. All rights reserved.

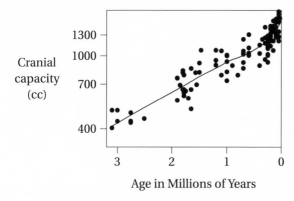

Figure 8.2: Cranial capacity of fossil hominins of, past 3 million the years. Each symbol represents a single skull. Drawn from data provided by Milford Wolpoff.

Although they could walk as we do, they were much better at climbing. Their finger bones were curved like those of apes, an adaptation that helps in swinging from overhead branches. They had short legs, long arms, and big feet. All of these are adaptations for climbing, and all betray the ancestry of these creatures as apes. These early hominins are intermediate fossils by any reasonable definition.

So far I have avoided naming species of fossil hominin. Experts disagree about which fossils belong to which species and even about the number of species. I wanted to make the case for intermediate forms without getting tangled up in these controversies. Yet the controversies are illuminating in their own way. In Figure 8.1, skulls older than 2 million years are called australopithecines and those younger than 1.8 million years are seen as members of our own genus, *Homo*. In between lies a troublesome group (skulls k–m) that has proved difficult to name. They have traditionally been called *Homo habilis*, but this name has never seemed entirely satisfactory. Some authors see these creatures as australopithecines; others see them as bona fide members of *Homo*. Lately, the consensus has swung toward the view that there are really two species here, *H. habilis* and another called *H. rudolfensis*. But that has not worked out too well either. As paleoanthropologist Ian Tattersall recently remarked, "just what—if anything—*Homo rudolfensis* is remains anybody's guess" [106]. It is interesting that these fossils are so hard to classify. In some features, they resemble the earlier forms. In others, they resemble later ones. They are, in other words, intermediate.

The impossibility of defining clear boundaries has also affected the way creationists discuss these fossils. These authors all argue that each hominin fossil is either "ape" or "human," with no intermediates. They do not however

Table 8.1: How different creationists classify hominin fossils into "ape" (A) and "human" (H).

	Creationist Publications						
Specimen	1	2	3	4–7	8–9	10–11	12
ER 1813	A	A	A	A	A	A	H
Java	A	A	H	A	A	H	H
Peking	A	A	H	A	H	H	H
ER 1470	A	A	A	H	H	H?	H?
ER 3733	A	H	H	H	H	H	H
WT 15000	A	H	H	H	H	H	H

From reference [41]. Key: 1, Cuozzo 1998; 2, Gish 1985; 3, Mehlert 1996; 4, Bowden 1981; 5, Menton 1988; 6, Taylor 1992; 7, Gish 1979; 8, Baker 1976; 9, Taylor & Van Bebber 1995; 10, Taylor 1996; 11, Lubenow 1992; 12, Line 2000.

agree about which is which (Table 8.1). Some authors even seem unable to make up their own minds. As James Foley (who compiled these data) observes, creationists

> assert that apes and humans are separated by a wide gap. If this is true, deciding on which side of that gap individual fossils lie should be trivially easy. Clearly, that is not the case. [41]

In summary, there is no shortage of intermediate forms in the hominin fossil record.

Genetics

If humans did evolve, then we share ancestors with other species, and that shared ancestry should be evident in our DNA. This section explores that evidence using methods introduced on pp. 25–31 of Chapter 3. The focus there was on whales rather than humans, but otherwise the problem is the same. We made use of genetic characters called transposons, which are often called "jumping genes." They "jump" by tricking the cell's machinery into copying them and inserting those copies at random in the DNA. As we have seen, these characters are ideal markers of common ancestry. All individuals who carry a given transposon share an ancestor, and essentially all the descendants of that ancestor carry the transposon.

To make sense of the human transposon data, we will need one additional idea, which I introduce by means of an evolutionary puzzle involving two transposons. Transposons do not have intuitively appealing names, and the two in question are called Ye5AH93 and Ye5AH137. The first is present in humans, chimpanzees, and bonobos (pygmy chimpanzees) but not in gorillas. As you know from Chapter 3, this pattern implies that humans, chimps, and bonobos share a common ancestor—one not shared with gorillas. The two chimpanzees, in other words, are our closest living relatives. The puzzle emerges when we examine the second transposon. It is found in humans and gorillas but not in chimps or bonobos. This implies that we are more closely related to gorillas than to chimps or bonobos. These cannot both be true. We cannot be closer to chimps than gorillas and *also* closer to gorillas than chimps. At least one of these transposons is lying.

At first glance, this conflict would appear to be a serious problem, undermining any effort to use transposons to argue for human evolution. Yet it would be far more surprising if such conflicts never occurred. Far from being a problem, they provide an opportunity to learn things about the past that we could never otherwise know.

To understand these wild claims of mine, consider what happens immediately after a species splits in two. The two daughter species are initially similar because many genetic variants are present in both. As time passes, each daughter will lose some genetic variants at random and gain others by mutation. Eventually, each daughter species will have its own unique suite of genetic variants—they will be perfectly sorted. But this sorting is a slow, gradual process. For many generations, the genetic material in the two daughter species will be incompletely sorted.

To see how this interacts with phylogeny, consider Figure 8.3. In the initial species there are two genetic variants, a mutant X and its ancestor O. The initial species splits and then splits again, forming three daughter species. Because the second split occurs before sorting is complete, both genetic variants are initially present in all three daughter species. These variants are neither advantageous nor harmful, so their frequencies drift around at random. Eventually, one will be lost in each species, but there is no telling which. In the figure, species A and C end up with variant X (the mutant form) and species B with variant O.

This result seems calculated to deceive some unwary geneticist. As explained on p. 25, we ordinarily expect species who share the mutant form of a character to lie on the same branch of the phylogenetic tree. But that is not what has happened here. Species A and B are on the same branch, yet it is A and C that share the mutant form of the character.

It is not that the sorting *had* to work out this way. For some other pair of genetic variants, the result might have been altogether different. This is why incomplete sorting can lead to conflict between characters. And this applies to *any* genetic character—even transposons. This process probably explains

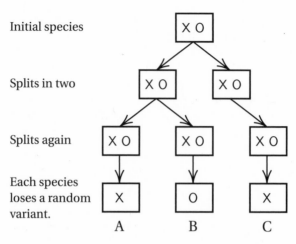

Initial species

Splits in two

Splits again

Each species
loses a random
variant.

A   B   C

Figure 8.3: Why incomplete sorting can make characters lie about phylogeny.

the conflict discussed above between Ye5AH93 and Ye5AH137. One of these characters is giving the wrong answer, but without more data we cannot tell which.

This is why biologists use lots of characters, allowing each one to "vote" for the phylogenetic tree of its choice. We gain confidence in a tree only when it receives support from many different characters. Fortunately, there is no shortage of transposons. Mark Batzer's lab at Louisiana State University has been studying them for years and has published a data set of 133 transposons typed in apes and humans [98]. Several had missing values like those in Figure 3.10 (p. 29). These were no problem for Batzer's group, for they analyzed the data using modern computer programs that take missing values in stride. But they would give *us* a headache, so we will study a reduced set of 73 transposons that excludes the ones with missing data and also the uninformative ones (found in just one species). These data are tabulated in Figure 8.4.

Recall that the whale transposons of Figure 3.8 (p. 27) fell into obvious "blocks." Each block consisted of several transposons with identical patterns of presence and absence across species. This block structure was our first clue of a phylogenetic pattern, and the same pattern is present in primate transposons. However, there are too many to fit onto a page, so Figure 8.4 uses a single column for each block of transposons.

Block A consists of a single transposon found only in chimpanzees and bonobos. It shows that these two species have a common ancestor not shared

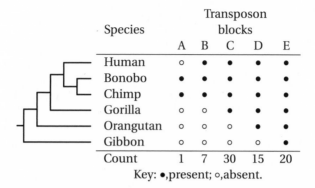

Species	Transposon blocks				
	A	B	C	D	E
Human	○	●	●	●	●
Bonobo	●	●	●	●	●
Chimp	●	●	●	●	●
Gorilla	○	○	●	●	●
Orangutan	○	○	○	●	●
Gibbon	○	○	○	○	●
Count	1	7	30	15	20

Key: ●,present; ○,absent.

Figure 8.4: Transposon data and tree for humans and apes [98]. Each column represents several transposons with identical presence/absence pattern. "Count" shows number of transposons in each block. These data exclude transposons with missing data and those present in just one.

with any other species. In other words, these two species lie on their own branch within the tree. Next we have block B, which represents 7 transposons shared uniquely by humans, bonobos, and chimps. It documents a branch including these three species. If you continue in this way, you will reconstruct the tree shown in the figure. This is the same tree that Batzer's group arrived at using all 133 transposons. As with the whale data, the block structure in these data is strong evidence that these species evolved from a common ancestor. There is no trace in Figure 8.4 of the character conflict described above. The troublesome transposon (Ye5AH137) was excluded because it contains a missing value. It turns out that in the whole data set of 133 transposons, this is the *only* one that conflicts with the tree in Figure 8.4.

It is interesting that only 1 transposon in 133 is in conflict. This tells us that the sorting process described in Figure 8.3 was nearly complete before the separation of humans and chimpanzees. Such evidence is often used to study the duration and the size of the species ancestral to humans and chimpanzees [102]. It is thus a source of information about the past, not a problem for the view that humans evolved.

Since nearly all the transposons vote for the same tree, the transposon data provide strong support for that tree. To see how strong that support really is, consider the creationist alternative. As discussed above, it is highly unlikely that two species would by chance have identical transposons at the same spot in their DNA. Yet several of the transposons in this data set are shared not just by two species but by several. On the creationist hypothesis, each of these transposons represents an improbability that verges on the miraculous. Yet our data contain dozens of them. Furthermore, these transposons are distributed among the species in a highly non-random fashion. We saw on p. 28 that 16 transposons, scattered randomly among species, are exceedingly unlikely to be consistent with any tree at all. Yet even the reduced data set in Figure 8.4 has 73 transposons, all of them consistent with a single tree. This requires another improbability of miraculous proportion. The creationist alternative is thus a matter of compounded miracles. One might argue that this is no problem because a deity can accomplish miracles. But if that is what happened, then I would submit that it was no accident. It must have been part of a deliberate effort to convince us that humans evolved.

And it was not the whole of that effort, for there is a great deal of other evidence. Our evolutionary past has left a trail of wreckage throughout our DNA. There are entire genes that no longer work. Take for example the enzyme *urate oxidase*. In most mammals, this enzyme assists in the chemical reaction that gets rid of uric acid. Humans lack this enzyme, and our blood is therefore high in uric acid. We have several times as much as most mammals. Our level is so high that we can easily push it beyond the safe limit. Gain a little weight, or increase your intake of beans or beer, and you may find yourself hobbling about with excruciating pain in the foot or ankle. This condition, which is called "gout", happens when uric acid becomes so concentrated that it begins to precipitate out of the blood, forming crystals in your joints.

Our high level of uric acid is not altogether bad, for the enzyme also functions as an antioxidant. In this role, it prevents damage to cells and may have something to do with our long lifespan. For this reason, it is plausible that a deity might have designed humans without urate oxidase. The trouble with this idea is that although we lack the enzyme, we nonetheless carry the gene that produces it. Ours is just broken. If the deity didn't want us to make the enzyme, why did he give us the gene?

Broken genes such as this one are called *pseudogenes*. Some creationists object that it is never possible to prove that a pseudogene is devoid of function. They produce no protein, but perhaps they have some unknown function. Although this is possible, it doesn't affect the conclusion. Suppose you saw someone tightening a screw with a pocket knife whose blade had been snapped in half. You would understand immediately that it had once been a pocket knife, even if it had lately been used as a screwdriver. It is the same with pseudogenes. Whatever function the human urate oxidase pseudogene may serve today, it was built to make enzyme. What is it doing in us?

This question would be easier for creationists if we were the only mammals that lacked urate oxidase, but we share this deficit with chimpanzees, gorillas, orangutans, and gibbons. To an evolutionist, this pattern suggests that these species all share a common ancestor. Note the consistency between this conclusion and the one we reached above. The species that are grouped together here also share 20 of the transposons tabulated in Figure 8.4. These transposons also imply that these same species share a common ancestor.

But let us take a closer look. Not only do these species lack urate oxidase, they also carry the urate oxidase pseudogene. Let us examine these pseudogenes closely to find out how they are broken. Our copy is broken in several places, one of which occurs at the gene's 12th amino acid. In most mammals, this amino acid is arginine and is produced by three adjacent nucleotides—CGA—in the DNA. In our DNA, the corresponding bit reads TGA. This tiny change has a large effect. It derails the cell's machinery and prevents it from producing any useful protein. Even if our gene were otherwise undamaged, this single change would have broken it. And what of the other species? We find precisely the same change at precisely the same spot in the DNA of chimpanzees, gorillas, and orangutans [82]. This is powerful support for the view that we share an ancestor with these species. The creationist alternative would be forced to assume that the deity had for some reason chosen to break the gene in precisely the same way in all these species. If he did so, then once again I submit that he did it on purpose.

As mentioned above, gibbons also lack urate oxidase. They also carry the urate oxidase pseudogene. Their copy however was broken in an entirely different way. They lack the enzyme because of an independent evolutionary event. This is analogous to the independent gene duplication discussed on p. 44, which allows howler monkeys to see in three-way color vision, much as monkeys and apes of the Old World do.

I have focused on just one pseudogene, but there are lots of others. One of them (GULOP) is the reason that our diets must include vitamin C. This gene is broken in precisely the same way in humans, apes, and African monkeys [71, pp. 97–99]. In another, whose nickname is ACYL3, the damage is shared only by humans and chimps [125]. This list could go on and on, and each pseudogene tells a similar story. In each one, we find damage that is shared by several species. In each case, the damage is so precisely identical that it could not plausibly have happened by chance.

So far, we have been looking at the DNA a little piece at a time: a transposon here, a pseudogene there. We can also look for traces of common ancestry in the DNA taken as a whole. This project is easier now than ever before, for geneticists have sequenced the entire DNA sequences of several primates. These data show that we differ from chimps at only a tiny fraction—1.23%—of our nucleotide sites [17]. At all the rest—about 3 billion sites in all—we are identical. And it isn't just humans and chimps. The DNA of humans and apes exhibits a perfect phylogenetic pattern [101]—exactly the same pattern implied by the transposons in Figure 8.4. This pattern is additional strong evidence that humans and apes evolved from a common ancestor.

Conclusion

The evidence for human evolution has become overwhelming. The fossil record is replete with intermediate forms, and evidence of common ancestry pervades our DNA. Similar evidence (which I have not discussed) pervades the disciplines of comparative anatomy and embryology. As Theodosius Dobzhansky once put it, "nothing in biology makes sense except in the light of evolution" [39]. This pattern could of course have been produced by a deity, even if humans were created separately. But not even a deity could have done this by accident. It would have required a deliberate effort to make it appear that humans evolved.

9

ARE WE STILL EVOLVING?

As recently as the year 2000, it was possible for Stephen Jay Gould to argue that

> Natural selection has almost become irrelevant in human evolution. There's been no biological change in humans in 40,000 or 50,000 years. [48]

This position seems easy to defend. We often respond to environmental challenges using culture. I wear eyeglasses to correct my vision. When it's cold, I wear a coat. Antibiotics protect me from disease. These and many other cultural responses reduce the force of natural selection. For this reason, many have argued that humans respond to the environment by changing their culture, not their genes.

On the other hand, culture also has an opposing effect. Not only does it buffer us from change, it also creates new challenges. We eat many foods that our ancestors got less of. Our dense cities increase the threat of infectious disease. These new conditions require new adaptations. One might even argue that culture has accelerated adaptive evolution.

A decade ago, when Gould gave the interview quoted above, there was no way to tell which position was correct. Gould seems to have favored the first, yet if he were still alive he might not do so today. That position has grown harder to defend because of a steady increase in knowledge about human genetics. When I started graduate school in the mid-1970s, geneticists described human variation at just a handful of positions within the DNA. Many studies used the same six. Nowadays, we routinely use over a million, and we answer questions that would once have seemed impossible.

One of these questions involves natural selection. To understand how this works, you need to imagine what happens when a favorable mutation first happens to some lucky chromosome. Being favorable, the new mutation has a better-than-average chance of spreading through the population. Naively, we might assume that this spread would involve the entire chromosome—that the lucky chromosome would simply increase in frequency until everyone carried two copies of it. Reality is slightly more complex.

The chromosome that you pass on to your son or daughter may not be identical to either of those you got from Mom and Dad. Instead, it is often a mosaic. It might start with a stretch of Dad's chromosome but finish with a stretch of Mom's. Or it might begin and end with Mom's but have a stretch of Dad's in the middle. In this sense, the chromosome you pass on to your child may have your father's eyes and your mother's hair. Geneticists call this *recombination.*

Because of recombination, chromosomes do not pass unscathed across the generations. Let us concentrate on the chromosomes that carry our favorable mutation—the lucky chromosomes. At first, there is just one of them. A few generations later, there may be several, but they will no longer be identical. Each time a lucky chromosome recombines with an unlucky one, it introduces something new. A segment of chromosome that had heretofore been unlucky gets spliced into a lucky chromosome. The favorable mutation will still sit on a region of DNA undisturbed by recombination, but over time this region will get smaller and smaller. This is the key that allows us to detect ongoing natural selection.

To see why, consider the case of a neutral mutation—one neither favorable nor unfavorable. Such a mutation has no inherent tendency to increase or decrease in frequency. From generation to generation, its frequency will change through accidents of sampling, but it is as likely to decrease as to increase. Over time, its frequency will wander up and down, and it may accidentally drift to a high frequency. This happens slowly, however, so there is plenty of time for recombination to work. Such a mutation will still sit on a region of undisturbed DNA, but that region will be small.

The opposite holds for favorable mutations. These tend to increase rapidly, pulled along by natural selection. Because they reach high frequencies quickly, there is less time for recombination, and this makes them easy to spot. We simply look for common alleles surrounded by large regions of undisturbed DNA.

But how can we recognize DNA that is undisturbed by recombination? That part is easy. In the undisturbed region, there should be two categories of chromosome: the lucky ones, which carry the favorable allele, and the rest. The lucky chromosomes should be similar to one another, for they are all copies of the same original chromosome, and there has not been much time for mutations to accumulate. Outside the undisturbed region, we expect more variation, and the same is true of the unlucky chromosomes within the undisturbed region. Thus, the undisturbed portions of the lucky chromosomes are distinctively different; they stand out like a black cat in the snow.

This pattern is obvious in Table 9.1, which shows genetic data from a region near the human lactase gene, typed in a sample of Europeans. The top row—called the reference sequence—shows the most common state at each nucleotide site. For example, most of the chromosomes in the sample had "a" in position 1. This value is shown in the first column of the reference sequence. Chromosomes that share this value have "." in that column.

Note the difference between the top and bottom halves of the table. The top half is nearly all dots because those chromosomes are nearly identical, both to each other and to the reference sequence. The bottom half of the table shows far more variation. This pattern suggests that the upper chromosomes have recently spread through the population by natural selection. We might be seeing a collection of those lucky chromosomes that I discussed above.

Before reaching such a conclusion, we need to ask another question. As we have seen, even neutral alleles sit on blocks of undisturbed DNA. But neutral alleles sit on small blocks and selected alleles on large ones. Thus, we need to ask about the *size* of the block of undisturbed DNA shown in Table 9.1. The table itself shows 84 nucleotide sites, but these are not adjacent on the chromosome. They are spread across more than 100,000 sites. This may sound like a lot, but it is by no means the entire block. I showed only a few sites because I had to fit the table onto a page. The entire block of undisturbed DNA occupies nearly a million nucleotide sites in the European population. It is one of the largest such blocks in human DNA. This provides strong evidence for natural selection.

The signature of selection is pretty obvious in Table 9.1, but that did not make it easy to find. We can see it only because someone scanned millions of variant sites within human DNA and put the data into a convenient table. Such data have become available only in the past few years, and we are just beginning to make sense of them.

These studies show, beyond any reasonable doubt, that natural selection has affected our species within the past few tens of thousands of years. There is a handful of genes, such as the lactase gene, at which evidence for selection is beyond dispute. Several studies have made an even stronger claim: that there

---

Table 9.1 *(following page)*: DNA sequences near the human lactase gene, typed in a European sample. Columns are nucleotide sites. The top row (the reference sequence) shows common state at each site. Capital "A" is the lactase persistence allele [8]. Dots indicate identity with top row. From HapMap [53] release r23.

```
gacattccgcttcaggcattcctatctaaacagaccaacgtaΛgggtacaatgcctaacccagacgtttcaactctggctgtta
```

```
..................................................................................
..................................................................................
..................................................................................
..................................................................................
..................................................................................
..................................................................................
..................................................................................
..................................................................................
..................................................................................
..................................................................................
..................................................................................
..................................................................................
..................................................................................
..................................................................................
..................................................................................
..................................................................................
..................................................................................
..................................................................................
..................................................................................
..................................................................................
..................................................................................
..................................................................................
..................................................................................
..................................................................................
..................................................................................
..................................................................................
..................................................................................
..................................................................................
..................................................................................
..................................................................................
..............................t...................................................
..............................t...................................................
.......................................................c..........................
.................................................g................................
ag.g.gt...........................................................................
..........................................G......................................
..........................................G......................................
..........................................G......................................
..........................................G......................................
..........................................G......................................
..........................................G..a.gt.....t.........gac.c.tgtct.....a..g
....c.....ccgga....gat..at..gg..c.....tc.gGaaa.g..ccttt...tg......c...t.t..........g
.g..c.....ccgga....gat..at..gg..c.....tc.gGaaa.g..ccttt...tg......c...t.t..........g
.gt.c.t..tcc...agtag.t.cat..g.....t..ttccgG..a.gt.....t.........gac.c.tgtct.........
....c.t..tcc...agtag.t.cat..g.....t.gttccgG..a.gt.....t.........gac.c.tgtct.....a..g
....c.t..tcc...agtag.t.cat..g.....t.gttccgG..a.gt.....t.........gac.c.tgtct.....a..g
.gt.c.t..tcc...agtag.t.cat..g.....t.g.tc.gG..a.gt.....t.........gac.c.tgtct........g
.gt.c.t..tcc...agtag.t.cat..g.....t..ttc.gG..acgt.....t.........gac.c.tgtct.....a..g
.gt.c.t..tcc...agtag.t.cat..g.....t.gttc.gG..a.gt.....t.........gac.c.tgtct.....a..g
.g..c.....ccgga....gat..at..gg..c.....tc.gGaaa.g..ccttt...tg......cg.gt.t..ctata.ccg
.gt.c.t..tcc...agtag.t.cat..g.....t.gttccgG..a.gt.....t.........gac.c.tgtct.....a..g
.gt.c.t..tcc...agtag.t.cat..g.....t.gttccgG..a.gt.....t.........gac.c.tgtct.....a..g
.gt.c.t..tcc...agtag.t.cat..g.....t.gttccgG..a.gt.....t.........gac.c.tgtct.....a..g
.g..c..tatccgga....g.tc.atcgg.tc.g.tg.tc.gG..a.g.g....tg....ggt...cg.gt.t..ctata.ccg
ag.g.gtta.ccgga....g.t..atcgg.tc.g.tg.tc.gG..a.g.g....tg....ggt...cg.gt.t..ct..a..g
ag.g.gtta.ccgga....g.t..atc.g.tc.g.tg.tc.gG..a.g.g....tg....ggt...cg.gt.t..ctata.ccg
ag.g.gtta.ccgga....g.t..atcgg.tc.g.tg.tc.gG..a.g.g....tg....ggt...cg.gt.t..ctata.ccg
ag.g.gtta.ccgga....g.t..atcgg.tc.g.tg.tc.gG..a.g.g....tg....ggt...cg.gt.t..ctata.ccg
ag.g.gtta.ccgga....g.t..atcgg.tc.g.tg.tc.gG..a.g.g....tg....ggt...cg.gt.t..ctata.ccg
ag.g.gtta.ccgga....g.t..atcgg.tc.g.tg.tc.gG..a.g.g....tg....ggt...cg.gt.t..ctata.ccg
ag.g.gtta.ccgga....g.t..atcgg.tc.g.tg.tc.gG..a.g.g....tg....ggtg..cg.gt.t..ctata.ccg
ag.g.gtta.ccgga....g.t..atcgg.tc.g.tg.tc.gG..a.g.g....tg....ggt...cg.gt.t..ctata.ccg
ag.g.gtta.ccgga....g.tc.atcgg.tc.g.tg.tc.gG..a.g.g....tg....ggt...cg.gt.t..ctata.ccg
ag.g.gtta.ccgga....g.t..atcgg.tc.g.tg.tc.gG..a.g.g....tg....ggtg..cg.gt.t..ctata.ccg
```

are hundreds—perhaps thousands—of such genes [111]. The human species cannot have been evolving at this rate for millions of years. If it had, the genetic difference between humans and chimpanzees would be much larger. These claims therefore imply that we are witnessing a spurt of adaptive evolution, which took place during the past few tens of thousands of years. If so, then human evolution did not stop when our species acquired culture. It did not even slow down. On the contrary, it *accelerated* [53]. These are strong claims and still controversial. There is no controversy, however, about the weaker claim that natural selection has affected *some* human genes during the past few tens of thousands of years.

The more we know about which genes have changed, the more important it becomes to understand what they do. In many cases, this is still mysterious. In others, at least part of the story is known. One of these is the lactase gene discussed above. Lactase is an enzyme that digests lactose, the sugar found in milk. Most mammals drink milk only as infants and stop making lactase shortly after they are weaned. Many humans do the same thing and are therefore unable to drink milk later in life. This condition is called *lactose intolerance* and is common throughout Asia and much of Africa. Yet many of us drink milk throughout life with no ill effects. This condition is called *lactase persistence* and is common in northern Europe and some parts of Africa.

The European form of lactase persistence results from a mutation close to the lactase gene, in the region responsible for turning the gene on and off. In effect, the mutation has disabled the gene's "off switch." This mutant allele is the one represented by a capital "A" in the central column of Table 9.1. It is apparently carried by all of the chromosomes in the top half of the table. Based on the length of the block of undisturbed DNA, Joel Hirschhorn and his colleagues estimate that it arose within the last 5,000–10,000 years and has increased rapidly in frequency ever since [8]. In some parts of northern Europe, it is at nearly 100%. This rapid rise in frequency left the dramatic pattern that is so obvious in the table.

In northern Europe we also find some of the earliest archaeological evidence of dairying. It appears that lactase persistence and dairying evolved together, the one by changes in genetics and the other by changes in culture. It makes sense that the two should have encouraged one another. Each increase in dairying made lactose persistence more valuable, and vice versa.

So far, genetics and archaeology are both consistent with the view that dairying and lactase persistence evolved together in Europe within the past 10,000 years. In addition, there is a third line of evidence that we can use to check the first two. If we had a time machine, we could travel back to the northern Europe of 6 or 8 thousand years ago and check prevailing levels of lactase persistence. If lactase persistence has evolved within the past few thousand years, then those ancient populations should have less of it. There are no time machines, but we do have the next best thing: many ancient Europeans have left us their bones. With skill and luck, it is possible to extract DNA from these.

Joachim Burger and his colleagues have recently studied the DNA of several ancient Europeans [13]. The most recent skeleton, which dates from about 1,500 years before present BP, carried a single copy of the lactase persistence allele. The other nine range in age roughly from 4,000–7,700 BP and come from various sites in northern and eastern Europe. Not one of these had even a single copy of the persistence allele. We should not infer that this allele was altogether lacking from those ancient populations. Even if it were present, we might still have happened to sample nine individuals, each with two copies of the non-persistence allele. But that would be highly unlikely—about 4 chances in 10 billion—if the persistence allele were as common as it is in Europe today. Taken together, these data prove that the persistence allele has spread through the population. They prove, in other words, that evolutionary change has happened within Europe during the last few thousand years.

Conclusion

In summary, there is no reason to think that human evolution stopped or even slowed during recent human history. It may in fact have accelerated. What of the argument that culture buffers us against the forces of evolutionary change? There must be some truth to it, but so far its effect has not shown up in genetic studies. We do however find genetic evidence of the opposing effect. Culture continues to pose fresh problems that genetic evolution must solve.

10 CONCLUSIONS

T he argument about evolution has changed a lot since the 19th century, and those changes have been evident in every chapter of this book. People used to dwell on the "fixity of species." Today, we worry about the drug-resistant bacteria and the pesticide-resistant insects that have evolved in response to our own chemical warfare. We no longer give much thought to the fixity of species.

People used to think that "blending" was a central feature of heredity. It was hard to understand how natural selection could make headway against it. This argument was hard to answer, and by the early decades of the 20th century many scientists viewed natural selection as discredited. But this concern melted away as the new science of genetics figured out how heredity really worked. This issue is now so uncontroversial that it did not even appear in the preceding chapters.

During the 19th century, the smart money held that the earth had not existed for the eons of time that evolution seemed to require. This concern also melted away, as geophysicists learned how to date rocks using radioactivity.

In the 19th century, there was every reason for evolutionists to worry about the fossil record. Where were all those intermediate fossils that the theory seemed to require? Critics still rail about this today, but their case rings hollow, for there is now no shortage of intermediate fossils.

Darwin was first led to thoughts about evolution by the geographic distribution of plants and animals. Yet in his day much of that distribution remained an enigma. As late as 1947, creationists such as Douglas Dewar could pose such questions as this:

> The Coecilians (legless, worm-like, burrowing amphibia) occur in America from Mexico to Peru, Tropical Africa and the East Indies. How did they come to be thus distributed? [37, p. 145]

These concerns disappeared as we began to understand continental drift and as progress in molecular genetics made it possible to infer reliable phylogenetic trees.

During the 19th century, there was no evidence that all living things had evolved from a single ancestor. Even Darwin allowed for the possibility of several origins of life. I doubt he would do so today, for molecular evidence now leaves little doubt.

In each of these cases, a creationist argument was discredited by scientific progress. Yet creationists have developed no new arguments to take their places. I know of only two exceptions. The philosopher Karl Popper once flirted with the idea that evolutionary theory might be circular [88, p. 241] but later changed his mind [89]. The creationists John Whitcomb and Henry Morris have claimed that evolution contradicts the second law of thermodynamics [117, p. 226]. But this hardly counts as an innovation, as it rests on a simple misunderstanding of that law [99]. Apart from these abortive efforts, I think it is true that all modern creationist arguments originated in the 1860s and 1870s. The modern versions often have new labels, such as "irreducible complexity" and "specified complexity," but this is just marketing. Beneath the new labels, these are the same arguments that Pritchard and Murphy made long ago (see pp. 38–39).

The case for evolution was much shakier in 1870 than today. One might well ask why so many Victorians found it convincing. By the same token, the case for evolution is now very strong, yet many remain unconvinced. This is worth wondering about too. In my view, both questions have the same answer. Without evolution, biology is a huge mass of unconnected facts. Evolution is a lens that makes those facts jump sharply into focus. The real power of evolution to convince lies in the huge mass of facts that it manages so simply to explain, a point that biologists have been making since 1867 [113, p. 1]. The more familiar one is with the facts of nature, the more convincing is the case for evolution. For this reason, virtually all modern biologists are evolutionists, yet they find it hard to convince a general public that lacks their broad knowledge.

It was the same in the 19th century. The first and most enthusiastic Darwinists were scholars who studied botany, zoology, or comparative anatomy. Many of Darwin's fiercest critics were mathematicians, astronomers, and engineers. The less one knew of the facts of biology, the easier it was to be critical.

Imagine the difficulty faced by a person deeply familiar with the facts of nature but also deeply committed to a literal interpretation of Genesis. One such person was Philip Henry Gosse, whom we met on p. 7. Gosse became enthralled by nature as a young man and studied it throughout life. He published book after book on a wide range of topics, including insects, birds, mammals, mollusks, sea anemones, and corals. In later life he was especially interested in aquatic creatures, and he invented the aquarium in order to study them. He was one of the most popular nature writers of the 19th century. Yet he was also a serious scientist and was in 1856 elected to the Royal Society—perhaps the most prestigious scientific society in the world.

Gosse would have been well aware of the idea of evolution, even before Darwin published his famous book. Geologists such as Charles Lyell had argued that the earth was very old, and the fossils of Great Britain seemed to back him up. A scientist such as Gosse could hardly fail to see that this theory explained much that was otherwise mysterious.

Yet it also seemed to contradict the Old Testament. Gosse was deeply religious and often preached at the local congregation of the Plymouth Brethren, a Protestant religious sect. His religious views were strict and dictated a literal reading of the bible. These views led Gosse into an intellectual impasse. His biblical views about creation seemed to contradict the evidence of geology and indeed that of his own eyes. In 1857 he embraced a desperate hypothesis, which he published in a book called *Omphalos* [47]. The book held that God had created the earth 6,000 years ago, just as the bible seemed to imply (see p. 71). But he created it complete with all the signs of an ancient evolutionary history. All the strata were there at the moment of creation, as were the fossils within them. The book's title is the Greek word for "navel" and refers to the book's central metaphor. Gosse observed that the human navel is a relic of birth. Yet the first human, Adam, was not born but created. Would Adam have had a navel? Gosse's answer was emphatic: Adam *must* have had a navel, for to lack one is to be incompletely human. In the same way, the newly created world would necessarily have all the signs of a prior existence.

This argument left Gosse's contemporaries with mouths agape. According to the *Westminster Review*, it was "too monstrous for belief" [63]. As religious people saw it, Gosse's idea involved God in an enormous deception. Gosse's friend Charles Kingsley remarked that he could not believe that "God had written on the rocks one enormous and superfluous lie" [18]. Not only did the book fail to convince Gosse's contemporaries, it also failed to sell. Eventually the publisher cut off the bindings and sold most of the copies as scrap paper. The book sank into obscurity, where it remained for over a century.

During the past few decades, interest in *Omphalos* has revived. The idea resurfaced in the creationist literature in 1961, and the book itself is now back in print both in paperback and cloth editions [117, pp. 232–233]. The *Omphalos* hypothesis is promoted on creationist websites. In view of this resurgence, it is important to be clear about what the hypothesis entails. Having just read my own book, you are in a good position to do this. The *Omphalos* hypothesis implies that God created thousands of feet of layered strata suggesting that the earth's rocks accumulated over billions of years. Not only do these strata contain fossils, they contain fossils that are deceptively distributed so as to suggest that modern plants and animals evolved from earlier forms. God even adjusted atomic isotopes in these rocks so they would trick modern scientists into thinking the earth is very old. He adjusted the rings of ancient corals so they would support this mistaken inference. Even the molecules in living things conspire to deceive: they seem to show in exquisite detail that all species are related and descend from a common ancestor that lived over a billion years ago. On top of all this, God equipped living things with all the genetic machinery required for evolution to happen. He made it possible for us to *watch* evolution happening, even though it did not happen in the past.

In the words of Rabbi Natan Slifkin, "if God went to enormous lengths to convince us that the world is billions of years old, who are we to disagree" [103, p. 167]?

ACKNOWLEDGEMENTS

I have been studying evolution for 30 years, but that research prepared me poorly for the job of writing this book. Modern scientists are narrow specialists, and I am no exception. I had a lot to learn. Nearly every chapter led me into some corner of the field that I did not know. Each chapter was thus its own research project, and I would never have succeeded without help. Nearly every page has been evaluated and corrected by some relevant specialist, and I am grateful to them all.

Specialists however may not notice when the prose is dense and tedious—they read that sort of thing all the time. Fortunately, I got help from friends and family and from the students of several college classes.

My greatest debt is to my wife and colleague, Elizabeth Cashdan, who read several versions of every chapter. If the book reads well, she deserves much of the credit.

I owe specific thanks to: Mark Batzer, Jake Enk, Donald Feener, Sabine Fuhrmann, John Hawks, Henry Harpending, Chad Huff, Mark Jeffreys, Lynn Jorde, Ed Levine, Dan-E. Nilsson, James O'Connell, Renee Pennington, Anya Plutynski, Jackie Rabb, Nala Rogers, Jon Seger, Sylvia Torti, Josh Trammell, and Milford Wolpoff.

BIBLIOGRAPHY

[1] Christoph Adami. Reducible complexity. *Science*, 321:61–63, 2006.

[2] Per Erik Ahlberg and Jennifer A. Clack. The axial skeleton of the Devonian tetrapod Ichthyostega. *Nature*, 437(7055):137–140, 2005.

[3] Per Erik Ahlberg and Jennifer A. Clack. A firm step from water to land. *Nature*, 440:747–749, 2006.

[4] Stevan J. Arnold, Michael E. Pfrender, and Adam G. Jones. The adaptive landscape as a conceptual bridge between micro- and macroevolution. *Genetica*, 112(1):9–32, 2001.

[5] Nicholas H. Barton and Brian Charlesworth. Genetic revolutions, founder effects, and speciation. *Annual Reviews in Ecology and Systematics*, 15(1):133–164, 1984.

[6] William Bateson. Interspecific sterility. *Nature*, 110:76, 1922.

[7] Michael Behe. *Darwin's Black Box: The Biochemical Challenge to Evolution*. Free Press, New York, 1996.

[8] Todd Bersaglieri, Pardis C. Sabeti, Nick Patterson, Trisha Vanderploeg, Steve F. Schaffner, Jared A. Drake, Matthew Rhodes, David E. Reich, and Joel N. Hirschhorn. Genetic signatures of strong recent positive selection at the lactase gene. *American Journal of Human Genetics*, 74:1111–1120, 2004.

[9] Peter T. Boag. The heritability of external morphology in Darwin's ground finches (*Geospiza*) on Isla Daphne Major, Galapagos. *Evolution*, 37(5):877–894, 1983.

[10] Peter T. Boag and Peter R. Grant. Intense natural selection in a population of Darwin's finches (*Geospizinae*) in the Galápagos. *Science*, 214(4516):82–85, 1981.

[11] Wilmot H. Bradley. Varves and climate of the Green River epoch. *USGS Professional Paper*, 158:87–110, 1929.

[12] Christina L. Burch and Lin Chao. Evolution by small steps and rugged landscapes in the RNA virus $\phi 6$. *Genetics*, 151:921–927, 1999.

[13] Joachim Burger, Martina Kirchner, Barbara Bramanti, Wolfgaang Haak, and Mark G. Thomas. Absence of the lactase-persistence-associated allele in early Neolithic Europeans. *Proceedings of the National Academy of Sciences, USA*, 104(10):3736, 2007.

[14] Sean B. Carroll, Jennifer K. Grenier, and Scott D. Weatherbee. *From DNA to Diversity: Molecular Genetics and the Evolution of Animal Design*. Blackwell, Oxford, 2nd edition, 2005.

[15] Anthony R. Cashmore, Jose A. Jarillo, Ying-Jie Wu, and Dongmei Liu. Cryptochromes: Blue light receptors for plants and animals. *Science*, 284(5415):760–765, 1999.

[16] Henry F. Chambers. The changing epidemiology of *Staphylococcus aureus*? *Emerging Infectious Diseases*, 7(2):178–182, 2001.

[17] Chimpanzee Sequencing and Analysis Consortium. Initial sequence of the chimpanzee genome and comparison with the human genome. *Nature*, 437(7055):69–87, 2005.

[18] Susan Chitty. *The Beast and the Monk: A Life of Charles Kingsley*. Hodder and Stoughton, London, 1974.

[19] Bryan Clarke and James Murray. Ecological genetics and speciation in land snails of the genus *Partula*. *Biological Journal of the Linnean Society*, 1:31–42, 1969.

[20] Mark T. Clementz, Anjali Goswami, Philip D. Gingerich, and Paul L. Koch. Isotopic records from early whales and sea cows: Contrasting patterns of ecological transition. *Journal of Vertebrate Paleontology*, 26(2): 355–370, 2006.

[21] James A. Cotton and Roderic D. M. Page. Going nuclear: Gene family evolution and vertebrate phylogeny reconciled. *Proceedings of the Royal Society of London, Series B*, 269:1555–1561, 2002.

[22] Jerry A. Coyne, Nicholas H. Barton, and Michael Turelli. Perspective: A critique of Wright's shifting balance theory of evolution. *Evolution*, 51(3):643–671, 1997.

[23] Jerry A. Coyne and H. Allen Orr. *Speciation*. Sinauer, Sunderland, Massachusetts, 2004.

[24] Georges Cuvier. *The Animal Kingdom*. Carvill, New York, 1832. Translated by H. M'Murtrie.

[25] E.B. Daeschler, N.H. Shubin, and F.A. Jenkins. A Devonian tetrapod-like fish and the evolution of the tetrapod body plan. *Nature*, 440(7085):757–763, 2006.

[26] G. Brent Dalrymple. *The Age of the Earth*. Stanford University Press, Stanford, California, 1991.

[27] Charles Darwin. *On the Origin of Species by Means of Natural Selection, or the Preservation of Favoured Races in the Struggle for Life*. John Murray, London, 1st edition, 1859.

[28] Charles Darwin. *The Descent of Man, and Selection in Relation to Sex*, Vol. 2. John Murray, London, 1871.

[29] Charles Darwin. *On the Origin of Species by Means of Natural Selection, or the Preservation of Favoured Races in the Struggle for Life*. John Murray, London, 6th edition, 1871.

[30] Charles R. Darwin. *Journal of Researches into the Natural History and Geology of the Countries Visited during the Voyage of H.M.S. Beagle Round the World, under the Command of Capt. Fitz Roy, R.N.* John Murray, London, 2nd edition, 1845.

[31] Richard Dawkins. *The Blind Watchmaker*. Norton, New York, 1986.

[32] Richard Dawkins. *Climbing Mount Improbable*. Norton, New York, 1997.

[33] Gavin de Beer. *Homology, An Unsolved Problem*. Oxford University Press, Oxford,. 1971.

[34] Michael Denton. *Evolution: A Theory in Crisis*. Alder and Alder, Chevy Chase, Maryland, 1986.

[35] Douglas Dewar. *Difficulties of the Evolution Theory*. E. Arnold and Co., London, 1931.

[36] Douglas Dewar, J.B.S. Haldane, and L. Merson Davies. *Is Evolution a Myth? A Debate between Douglas Dewar, L. Merson Davies and J.B.S. Haldane*. C.A. Watts/Paternoster Press, London, 1949.

[37] Douglas Dewar and H. S. Shelton. *Is Evolution Proved? A Debate between Douglas Dewar and H.S. Shelton*. Hollis and Carter, London, 1947.

[38] Lettice Digby. The cytology of *Primula kewensis* and of other related primula hybrids. *Annals of Botany*, 36:357–388, 1912.

[39] Theodosius Dobzhansky. Nothing in biology makes sense except in the light of evolution. *The American Biology Teacher*, 35:125–129, 1973.

[40] Millard J. Erickson. *Christian Theology*, Vol. 1. Baker Book House, Grand Rapids, Michigan, 1983.

[41] James Foley. Comparison of all skulls. http://www.talkorigins.org/faqs/homs/compare.html.

[42] Walter Gehring and Kazuho Ikeo. Pax 6: Mastering eye morphology and eye evolution. *Trends in Genetics*, 15(9):371–377, 1999.

[43] D.N. Gilbert, S.J. Kohlhepp, K.A. Slama, G. Grunkemeier, G. Lewis, R.J. Dworkin, S.E. Slaughter, and J.E. Leggett. Phenotypic resistance of *Staphylococcus aureus*, selected *Enterobacteriaceae*, and *Pseudomonas aeruginosa* after single and multiple in vitro exposures to ciprofloxacin, levofloxacin, and trovafloxacin. *Antimicrobial Agents and Chemotherapy*, 45(3):883–892, 2001.

[44] Philip D. Gingerich, S. Mahmood Raza, Muhammad Arif, Mohammad Anwar, and Xiaoyuan Zhou. New whale from the Eocene of Pakistan and the origin of cetacean swimming. *Nature*, 368:844–847, 1994.

[45] Philip D. Gingerich, Munir ul Haq, Iyad S. Zalmout, Intizar Hussain Khan, and M. Sadiq Mulkani. Origin of whales from early artiodactyls: Hands and feet of Eocene protocetidae from Pakistan. *Science*, 293:2239–2242, 2001.

[46] Edmund Gittenberger. Sympatric speciation in snails; a largely neglected model. *Evolution*, 42(4):826–828, 1988.

[47] Philip H. Gosse. *Omphalos: An Attempt to Untie the Geological Knot*. John Van Voorst, London, 1857.

[48] Stephen Jay Gould. The spice of life: An interview with Stephen Jay Gould. *Leader to Leader*, Winter 2000.

[49] Peter R. Grant and B. Rosemary Grant. Hybridization of bird species. *Science*, 256(5054):193–197, 1992.

[50] Peter R. Grant and B. Rosemary Grant. Evolution of character displacement in Darwin's finches. *Science*, 313:224–226, 2006.

[51] Peter R. Grant and B. Rosemary Grant. *How and Why Species Multiply: The Radiation of Darwin's Finches*. Princeton University Press, Princeton, New Jersey, 2008.

[52] Verne Grant. *Plant Speciation*. Columbia University Press, New York, 2nd edition, 1981.

[53] John Hawks, Eric T. Wang, Grigory M. Cochran, Henry C. Harpending, and Robert K. Moyzis. Recent acceleration of human adaptive evolution. *Proceedings of the National Academy of Sciences, USA*, 104:20753–20758, 2008.

[54] Gertrude Himmelfarb. *Darwin and the Darwinian Revolution*. Norton, New York, 2nd edition, 1962.

[55] Keiichi Hiramatsu, Longzhu Cui, Makoto Kuroda, and Teruyo Ito. The emergence and evolution of methicillin-resistant *Staphylococcus aureus*. *Trends in Microbiology*, 9(10):486–493, 2001.

[56] D.L. Hull. *Darwin and his Critics*. Harvard University Press, Cambridge, Massachusetts, 1973.

[57] Thomas H. Huxley. A critical examination of the position of Mr. Darwin's work "On the Origin of Species," in relation to the complete theory of the causes of the phenomena of organic nature. In *On Our Knowledge of the Causes of the Phenomena of Organic Nature*, chapter VI, pp. 132–156. Robert Hardwicke, London, 1863.

[58] Darren E. Irwin, Jessica H. Irwin, and Trevor D. Price. Ring species as bridges between microevolution and speciation. *Genetica*, 112–113(1): 223–243, 2001.

[59] Fleeming Jenkin. The origin of species. In David L. Hull, editor, *Darwin and his Critics*, pp. 303–350. Harvard University Press, Cambridge, Massachusetts, 1973.

[60] Davida Kellogg and James D. Hays. Microevolutionary patterns in late Cenozoic Radiolaria. *Paleobiology*, 1(2):150–160, 1975.

[61] H. Kitagawa and J. van der Plicht. Atmospheric radiocarbon calibration to 45,000 yr B.P.: Late glacial fluctuations and cosmogenic isotope production. *Science*, 279:1187–1190, 1998.

[62] Alois F. Kovarik and Norman I. Adams, Jr. A new determination of the disintegration constant of uranium by the method of counting α–particles. *Physical Review*, 40:718–726, 1932.

[63] D. Krause. Apparent age and its reception in the 19th century. *Journal of the American Scientific Affiliation*, 32(3):146–150, 1980.

[64] Bernd Kromer. Revision and tentative extension of the tree-ring based ^{14}C calibration, 9200–11,855 CAL BP. *Radiocarbon*, 40(3):1117–1125, 1998.

[65] A.M. Labin and E.N. Ribak. Retinal glial cells enhance human vision acuity. *Physical Review Letters*, 104(15):158102, 2010.

[66] Philip Lake. Origin of continents. In *Encyclopædia Britannica*, Vol. 6, pp. 335–337. Encyclopædia Britannica Inc., London, 1960.

[67] Michael F. Land and Russell D. Fernald. The evolution of eyes. *Annual Reviews in Neuroscience*, 15:1–29, 1992.

[68] Harlan Lewis. Speciation in flowering plants. *Science*, 152:167–172, 1966.

[69] Tamra C. Mendelson and Kerry L. Shaw. Rapid speciation in an arthropod. *Nature*, 433(7024):375–376, 2005.

[70] Kenneth R. Miller. The flagellum unspun: The collapse of "irreducible complexity." In William A. Dembski and Michael Ruse, editors, *Debating Design: From Darwin to DNA*, pp. 81–97. Cambridge University Press, Cambridge, 2004.

[71] Kenneth R. Miller. *Only a Theory: Evolution and the Battle for America's Soul*. Viking, New York, 2008.

[72] St. George Mivart. Difficulties of the theory of natural selection. *The Month*, 11:35–53, 134–153, 274–289, 1869. Published anonymously.

[73] Arne Müntzing. Cyto-genetic investigations on synthetic *Galeopsis tetrahit*. *Heriditas*, 16:105–154, 1932.

[74] Joseph John Murphy. Presidential address to the Belfast Natural History and Philosophical Society. *Northern Whig*, 19 November 1866.

[75] Joseph John Murphy. *Habit and Intelligence*, Vol. I. MacMillan, London, 1869.

[76] V.R. Murthy and C.C. Patterson. Primary isochron of zero age for meteorites and the earth. *Journal of Geophysical Research*, 67:1161–1167, 1962.

[77] Masato Nikaido, Fumio Matsuno, Healy Hamilton, Robert L. Brownell, Jr., Ying Cao, Wang Ding, Zhu Zuoyan, Andrew M. Shedlock, R. Ewan Fordyce, Masami Hasegawa, and Norihiro Okada. Retroposon analysis of major cetacean lineages: The monophyly of toothed whales and the paraphyly of river dolphins. *Proceedings of the National Academy of Sciences, USA*, 98(13):7384–7389, 2001.

[78] Masato Nikaido, Alejandro P. Rooney, and Norihiro Okada. Phylogenetic relationships among cetartiodactyls based on insertions of short and long interpersed elements: Hippopotamuses are the closest extant relatives of whales. *Proceedings of the National Academy of Sciences, USA*, 96(18):10261–10266, 1999.

[79] Dan-E. Nilsson and Susanne Pelger. A pessimistic estimate of the time required for an eye to evolve. *Proceedings of the Royal Society of London, Series B*, 256(1345):53–58, 1994.

[80] Dan-E. Nilsson. Vision optics and evolution. *BioScience*, 39(5):298–307, 1989.

[81] S. Nummela, J.G.M. Thewissen, S. Bajpai, S.T. Hussain, and K. Kumar. Eocene evolution of whale hearing. *Nature*, 430(7001):776–778, 2004.

[82] M. Oda, Y. Satta, O. Takenaka, and N. Takahata. Loss of urate oxidase activity in hominoids and its evolutionary implications. *Molecular Biology and Evolution*, 19(5):640–653, 2002.

[83] Brendan O'Fallon and Fredrick R. Adler. Stochasticity, complex spatial structure, and the feasibility of the shifting balance theory. *Evolution*, 60(3):448–459, 2006.

[84] D. Osorio. Eye evolution: Darwin's shudder stilled. *Trends in Ecology and Evolution*, 9(7):241–242, 1994.

[85] Daniel Otte. *The Crickets of Hawaii: Origin, Systematics and Evolution*. The Orthopterists' Society at the Academy of Natural Sciences of Philadelphia, 1994.

[86] William Paley. *Natural Theology, or, Evidences of the Existence and Attributes of the Deity*. Gould and Lincoln, Boston, 1860. Reprint of 1809 edition.

[87] Raymond Pearl. Data on the relative conspicuousness of barred and self-colored fowls. *American Naturalist*, 45(530):107–117, 1911.

[88] Karl R. Popper. *Objective Knowledge: An Evolutionary Approach*. Oxford University Press, Oxford, 1972.

[89] K.R. Popper. Natural selection and the emergence of mind. In *Evolutionary Epistemology, Rationality, and the Sociology of Knowledge*, pp. 139–155, Open Court Publishing, La Salle, Illinois, 1987.

[90] Charles Pritchard. *The Continuity of the Schemes of Nature and of Revelation. A Sermon Preached, by Request, on the Occasion of the Meeting of the British Association at Nottingham. With Remarks on some Relations of Modern Knowledge to Theology*. Bell and Daldy, London, 1866.

[91] P.R. Renne, W.D. Sharp, A.L. Deino, G. Orsi, and L. Civetta. ^{40}Ar/^{39}Ar dating into the historical realm: Calibration against Pliny the Younger. *Science*, 277:1279–1280, 1997.

[92] Paul R. Renne, Alan L. Deino, Robert C. Walter, Brent D. Turrin, Carl C. Swisher, Tim A. Becker, Garniss H. Curtis, Warren D. Sharp, and Abdur-Rahim Jaouni. Intercalibration of astronomical and radioisotopic time. *Geology*, 22(9):783–786, 1994.

[93] William R. Rice and Ellen E. Hostert. Laboratory experiments on speciation: What have we learned in 40 years? *Evolution*, 47:1637–1653, 1993.

[94] William R. Rice and George W. Salt. Speciation via disruptive selection on habitat preference: Experimental evidence. *American Naturalist*, 131:911–917, 1988.

[95] G.K. Roderick and R.G. Gillespie. Speciation and phylogeography of Hawaiian terrestrial arthropods. *Molecular Ecology*, 7(4):519–531, 1998.

[96] Paul M. Romer. Economic growth. In David R. Henderson, editor, *The Concise Encyclopedia of Economics*. The Library of Economics and Liberty, 2007.

[97] Ernest Rutherford and Frederick Soddy. The cause and nature of radioactivity. *Philosophical Magazine*, 4:370–396, 1902.

[98] Abdel-Halim Salem, David A. Ray, Jinchuan Xing, Pauline A. Callinan, Jeremy S. Myers, Dale J. Hedges, Randall K. Garber, David J. Witherspoon, Lynn B. Jorde, and Mark A. Batzer. *Alu* elements and hominid phylogenetics. *Proceedings of the National Academy of Sciences, USA*, 100(22):12787–12791, 2003.

[99] N. Shanks. *God, the Devil, and Darwin: A Critique of Intelligent Design Theory*. Oxford University Press, Oxford, 2004.

[100] Neil Shubin. *Your Inner Fish: A Journey into the 3.5-Billion-Year History of the Human Body*. Pantheon, New York, 2008.

[101] C.G. Sibley and J.E. Ahlquist. DNA hybridization evidence of hominoid phylogeny: Results from an expanded data set. *Journal of Molecular Evolution*, 26:99–121, 1987.

[102] Adam Siepel. Phylogenomics of primates and their ancestral populations. *Genome Research*, 19(11):1929–1941, 2009.

[103] Natan Slifkin. *The Challenge of Creation: Judaism's Encounter with Science, Cosmology, and Evolution*. Yashar Books, Brooklyn, New York, 2006.

[104] Frank J. Sulloway. Darwin's conversion: The Beagle voyage and its after-math. *Journal of the History of Biology*, 15(3):325–396, 1982.

[105] Alison K. Surridge, Daliel Osorio, and Nicholas I. Mundy. Evolution and selection of trichromatic vision in primates. *Trends in Ecology and Evolution*, 18(4):198–205, 2003.

[106] Ian Tattersall. *The Fossil Trail: How We Know What We Think We Know about Human Evolution*. Oxford University Press. Oxford, 2nd edition, 2009.

[107] J.G.M. Thewissen and Sunil Bajpai. Whale origins as a poster child for macroevolution. *BioScience*, 51(12):1037–1049, 2001.

[108] William Thomson. On the secular cooling of the earth. In *Mathematical and Physical Papers*, Vol. 3, pp. 295–311. C. J. Clay and Sons, 1890. Reprinted from *Transactions of the Royal Society of Edinburgh*, 23:157–170, 1862.

[109] James Ussher. *The Annals of the World*. E. Tyler, London, 1658.

[110] Sara Via. Reproductive isolation between sympatric races of pea aphids: I. gene flow restriction and habitat choice. *Evolution*, 53(5):1446–1457, 1999.

[111] Benjamin F. Voight, Sridhar Kudaravalli, Xiaoquan Wen, and Jonathan K. Pritchard. A map of recent positive selection in the human genome. *Public Library of Science, Biology*, 4(3):e72, 2006.

[112] L. von Salvini-Plawen and Ernst Mayr. On the evolution of photoreceptors and eyes. In Max K. Hecht, William C. Steere, and Bruce Wallace, editors, *Evolutionary Biology*, Vol. 10, chapter 4, pp. 207–263. Plenum Press, New York, 1977.

[113] Alfred Russel Wallace. Mimicry, and other protective resemblances among animals. *Westminster and Foreign Quarterly Review*, 88(173):1–43, July 1867.

[114] G.J. Wasserburg, A.L. Albee, and M.A. Lanphere. Migration of radiogenic strontium during metamorphism. *Journal of Geophysical Research*, 69:4395–4401, 1964.

[115] Jonathan Weiner. *The Beak of the Finch*. Vintage Books, New York, 1994.

[116] J.W. Wells. Coral growth and geochronometry. *Nature*, 197(4871):948–950, 1963.

[117] John C. Whitcomb and Henry M. Morris. *The Genesis Flood*. Baker Book House, Grand Rapids, Michigan, 1961.

[118] Michael C. Whitlock. Variance-induced peak shifts. *Evolution*, 49(2):252–259, 1995.

[119] Michael C. Whitlock. Founder effects and peak shifts without genetic drift: Adaptive peak shifts occur easily when environments fluctuate slightly. *Evolution*, 51(4):1044–1048, 1997.

[120] Pat Shipman. *Taking Wing. Archaeopteryx and the Evolution of Bird Flight*. Simon and Schuster, New York, 1998.

[121] Samuel Wilberforce. On *The Origin of Species. Quarterly Review*, 108:225–264, 1860.

[122] Simon A. Wilde, John W. Valley, William H. Peck, and Colin M. Graham. Evidence from detrital zircons for the existence of continental crust and oceans on the earth 4.4 Gyr ago. *Nature*, 409:175–178, 2001.

[123] George C. Williams. *Plan and Purpose in Nature*. Weidenfeld and Nicolson, London, 1996.

[124] George C. Williams. *The Pony Fish's Glow: And Other Clues to Plan and Purpose in Nature*. Basic Books, New York, 1997.

[125] J. Zhu, J.Z. Sanborn, M. Diekhans, C.B. Lowe, T.H. Pringle, and D. Haussler. Comparative genomics search for losses of long-established genes on the human lineage. *PLoS Computational Biology*, 3:e247, 2007.

[126] Carl Zimmer. *At the Water's Edge: Fish with Fingers, Whales with Legs, and How Life Came Ashore and then Went Back to Sea*. Simon and Schuster, New York, 1998.

INDEX